C0-AWT-905

A Short Introduction to Perturbation Theory for Linear Operators

Tosio Kato

A Short Introduction to Perturbation Theory for Linear Operators

Springer-Verlag
New York Heidelberg Berlin

Tosio Kato
Department of Mathematics
University of California
Berkeley, CA 94720
USA

AMS Subject Classifications (1980): 46-01, 34-01, 34D10, 47-01, 47A55

Library of Congress Cataloging in Publication Data

Kato, Tosio, 1917–
A short introduction to perturbation theory for linear operators
"Slightly expanded reproduction of the first two chapters (plus introduction) of...
Perturbation theory for linear operators" – Pref.
Bibliography: p.
Includes index.

1. Linear operators. 2. Perturbation
(Mathematics) I. Title
QA329.2.K3725 1982 515.7′246 82-10505

This book is a slightly expanded version of Chapters 1 and 2 of:
T. Kato, *Perturbation Theory for Linear Operators* 2nd ed.
(Grundlehren der mathematischen Wissenschaften 132).
Springer-Verlag, Berlin, 1980.

Typesetting by Brühlsche Universitätsdruckerei (FRG).
Printing and binding by R. R. Donnelley & Sons, Harrisonburg, VA.
Printed in the United States of America.

9 8 7 6 5 4 3 2 1

ISBN 0-387-90666-5 Springer-Verlag New York Heidelberg Berlin
ISBN 3-540-90666-5 Springer-Verlag Berlin Heidelberg New York

Preface

This book is a slightly expanded reproduction of the first two chapters (plus Introduction) of my book *Perturbation Theory for Linear Operators*, Grundlehren der mathematischen Wissenschaften 132, Springer 1980. Ever since, or even before, the publication of the latter, there have been suggestions about separating the first two chapters into a single volume. I have now agreed to follow the suggestions, hoping that it will make the book available to a wider audience.

Those two chapters were intended from the outset to be a comprehensive presentation of those parts of perturbation theory that can be treated without the topological complications of infinite-dimensional spaces. In fact, many essential and even advanced results in the theory have nontrivial contents in finite-dimensional spaces, although one should not forget that some parts of the theory, such as those pertaining to scattering, are peculiar to infinite dimensions.

I hope that this book may also be used as an introduction to linear algebra. I believe that the analytic approach based on a systematic use of complex functions, by way of the resolvent theory, must have a strong appeal to students of analysis or applied mathematics, who are usually familiar with such analytic tools.

In addition to minor local improvements and modifications throughout the two chapters, the following new sections and paragraphs have been added in the new version: in Chapter One, § 4.7 on product formulas, § 6.11 on dissipative operators and contraction semigroups, and § 7 on positive matrices; in Chapter Two, §§ 4.7–4.9 on the extended treatment of analytic perturbation theory, § 6.6 on nonsymmetric perturbation of symmetric operators, and § 7 on perturbation of (essentially) nonnegative matrices.

The numbering of chapters, sections, paragraphs, theorems, lemmas, etc. remains the same as in the larger volume. Thus "I-§ 2.3" denotes the third paragraph of the second section of Chapter One, and "Lemma I-2.3" is Lemma 2.3 of Chapter One; the chapter number is omitted when referred to within the same chapter. Due to the particular genesis of the new version, however, some irregularities occur in the numbering of newly introduced theorems, lemmas, etc. Thus, for example, II-§ 5.7 contains new Theorems 5.13a–5.13c and a new Remark 5.13d.

References are given by numbers in square brackets, such as ⟦1⟧ for books and [1] for monographs. They retain the same numbers as in the larger volume (unless they are new), although only those referred to in the present volume are listed in the Bibliography. Thus, for example, only the papers [1, 2, 3, 6, 9, 12, 13] are listed under T. Kato.

Another irregularity is that in a few places, particularly in the Introduction, there remain references to the later chapters not contained in the present version. But I thought it more convenient to retain, rather than eliminate, such formal imperfections.

I would like to thank the staff of Springer-Verlag for their encouragement and help in conceiving this new book.

Berkeley
January, 1982

TOSIO KATO

Contents

Chapter Two

Introduction

Throughout this book, "perturbation theory" means "perturbation theory for linear operators". There are other disciplines in mathematics called perturbation theory, such as the ones in analytical dynamics (celestial mechanics) and in nonlinear oscillation theory. All of them are based on the idea of studying a system deviating slightly from a simple ideal system for which the complete solution of the problem under consideration is known; but the problems they treat and the tools they use are quite different. The theory for linear operators as developed below is essentially independent of other perturbation theories.

Perturbation theory was created by RAYLEIGH and SCHRÖDINGER (cf. SZ.-NAGY [1]). RAYLEIGH gave a formula for computing the natural frequencies and modes of a vibrating system deviating slightly from a simpler system which admits a complete determination of the frequencies and modes (see RAYLEIGH ⟦1⟧, §§ 90, 91). Mathematically speaking, the method is equivalent to an approximate solution of the eigenvalue problem for a linear operator slightly different from a simpler operator for which the problem is completely solved. SCHRÖDINGER developed a similar method, with more generality and systematization, for the eigenvalue problems that appear in quantum mechanics (see SCHRÖDINGER ⟦1⟧, [1]).

These pioneering works were, however, quite formal and mathematically incomplete. It was tacitly assumed that the eigenvalues and eigenvectors (or eigenfunctions) admit series expansions in the small parameter that measures the deviation of the "perturbed" operator from the "unperturbed" one; no attempts were made to prove that the series converge.

It was in a series of papers by RELLICH that the question of convergence was finally settled (see RELLICH [1]–[5]; there were some attempts at the convergence proof prior to RELLICH, but they were not conclusive; see e. g. WILSON [1]). The basic results of RELLICH, which will be described in greater detail in Chapters II and VII, may be stated in the following way. Let $T(\varkappa)$ be a bounded selfadjoint operator in a Hilbert space H, depending on a real parameter $\varkappa$ as a convergent power series

$$T(\varkappa) = T + \varkappa\, T^{(1)} + \varkappa^2\, T^{(2)} + \cdots . \tag{1}$$

Suppose that the unperturbed operator $T = T(0)$ has an isolated eigenvalue λ (isolated from the rest of the spectrum) with a finite multiplicity m. Then $T(\varkappa)$ has exactly m eigenvalues $\mu_j(\varkappa)$, $j = 1, \ldots, m$

(multiple eigenvalues counted repeatedly) in the neighborhood of λ for sufficiently small $|\varkappa|$, and these eigenvalues can be expanded into convergent series

$$(2) \qquad \mu_j(\varkappa) = \lambda + \varkappa\, \mu_j^{(1)} + \varkappa^2\, \mu_j^{(2)} + \cdots, \quad j = 1, \ldots, m\,.$$

The associated eigenvectors $\varphi_j(\varkappa)$ of $T(\varkappa)$ can also be chosen as convergent series

$$(3) \qquad \varphi_j(\varkappa) = \varphi_j + \varkappa\, \varphi_j^{(1)} + \varkappa^2\, \varphi_j^{(2)} + \cdots, \quad j = 1, \ldots, m\,,$$

satisfying the orthonormality conditions

$$(4) \qquad (\varphi_j(\varkappa), \varphi_k(\varkappa)) = \delta_{jk}\,,$$

where the φ_j form an orthonormal family of eigenvectors of T for the eigenvalue λ.

These results are exactly what were anticipated by RAYLEIGH, SCHRÖDINGER and other authors, but to prove them is by no mean simple. Even in the case in which H is finite-dimensional, so that the eigenvalue problem can be dealt with algebraically, the proof is not at all trivial. In this case it is obvious that the $\mu_j(\varkappa)$ are branches of algebroidal functions of $\varkappa$, but the possibility that they have a branch point at $\varkappa = 0$ can be eliminated only by using the selfadjointness of $T(\varkappa)$. In fact, the eigenvalues of a selfadjoint operator are real, but a function which is a power series in some fractional power $\varkappa^{1/p}$ of $\varkappa$ cannot be real for both positive and negative values of $\varkappa$, unless the series reduces to a power series in $\varkappa$. To prove the existence of eigenvectors satisfying (3) and (4) is much less simple and requires a deeper analysis.

Actually RELLICH considered a more general case in which $T(\varkappa)$ is an unbounded operator; then the series (1) required new interpretations, which form a substantial part of the theory. Many other problems related to the one above were investigated by RELLICH, such as estimates for the convergence radii, error estimates, simultaneous consideration of all the eigenvalues and eigenvectors and the ensuing question of uniformity, and non-analytic perturbations.

RELLICH's fundamental work stimulated further studies on similar and related problems in the theory of linear operators. One new development was the creation by FRIEDRICHS of the perturbation theory of continuous spectra (see FRIEDRICHS [2]), which proved extremely important in scattering theory and in quantum field theory. Here an entirely new method had to be developed, for the continuous spectrum is quite different in character from the discrete spectrum. The main problem dealt with in FRIEDRICHS's theory is the similarity of $T(\varkappa)$ to T, that is, the existence of a non-singular operator $W(\varkappa)$ such that $T(\varkappa) = W(\varkappa)\, T\, W(\varkappa)^{-1}$.

The original results of RELLICH on the perturbation of isolated eigenvalues were also generalized. It was found that the analytic theory gains in generality as well as in simplicity by allowing the parameter $\varkappa$ to be complex, a natural idea when analyticity is involved. However, one must then abandon the assumption that $T(\varkappa)$ is selfadjoint for all $\varkappa$, for an operator $T(\varkappa)$ depending on $\varkappa$ analytically cannot in general be selfadjoint for all $\varkappa$ of a complex domain, though it may be selfadjoint for all real $\varkappa$, say. This leads to the formulation of results for non-selfadjoint operators and for operators in Banach spaces, in which the use of complex function theory prevails (SZ.-NAGY [2], WOLF [1], T. KATO [6]). It turns out that the basic results of RELLICH for selfadjoint operators follow from the general theory in a simple way.

On the other hand, it was recognized (TITCHMARSH [1], [2], T. KATO [1]) that there are cases in which the formal power series like (2) or (3) diverge or even have only a finite number of significant terms, and yet approximate the quantities $\mu_j(\varkappa)$ or $\varphi_j(\varkappa)$ in the sense of asymptotic expansion. Many examples, previously intractable, were found to lie within the sway of the resulting asymptotic theory, which is closely related to the singular perturbation theory in differential equations.

Other non-analytic developments led to the perturbation theory of spectra in general and to stability theorems for various spectral properties of operators, one of the culminating results being the index theorem (see GOHBERG and KREIN [1]).

Meanwhile, perturbation theory for one-parameter semigroups of operators was developed by HILLE and PHILLIPS (see PHILLIPS [1], HILLE and PHILLIPS 〚1〛). It is a generalization of, as well as a mathematical foundation for, the so-called time-dependent perturbation theory familiar in quantum mechanics. It is also related to time-dependent scattering theory, which is in turn closely connected with the perturbation of continuous spectra. Scattering theory is one of the subjects in perturbation theory most actively studied at present.

It is evident from this brief review that perturbation theory is not a sharply-defined discipline. While it incorporates a good deal of the spectral theory of operators, it is a body of knowledge unified more by its method of approach than by any clear-cut demarcation of its province. The underpinnings of the theory lie in linear functional analysis, and an appreciable part of the volume is devoted to supplying them. The subjects mentioned above, together with some others, occupy the remainder.

Chapter One

Operator theory in finite-dimensional vector spaces

This chapter is preliminary to the following one where perturbation theory for linear operators in a finite-dimensional space is presented. We assume that the reader is more or less familiar with elementary notions of linear algebra. In the beginning sections we collect fundamental results on linear algebra, mostly without proof, for the convenience of later reference. The notions related to normed vector spaces and analysis with vectors and operators (convergence of vectors and operators, vector-valued and operator-valued functions, etc.) are discussed in somewhat more detail. The eigenvalue problem is dealt with more completely, since this will be one of the main subjects in perturbation theory. The approach to the eigenvalue problem is analytic rather than algebraic, depending on function-theoretical treatment of the resolvents. It is believed that this is a most natural approach in view of the intended extension of the method to the infinite-dimensional case in later chapters.

Although the material as well as the method of this chapter is quite elementary, there are some results which do not seem to have been formally published elsewhere (an example is the results on pairs of projections given in §§ 4.6 and 6.8).

§ 1. Vector spaces and normed vector spaces

1. Basic notions

We collect here basic facts on finite-dimensional vector spaces, mostly without proof[1]. A *vector space* X is an aggregate of elements, called *vectors*, $u, v, \ldots$, for which *linear operations* (addition $u + v$ of two vectors u, v and multiplication αu of a vector u by a *scalar* α) are defined and obey the usual rules of such operations. Throughout the book, the scalars are assumed to be *complex numbers* unless otherwise stated (complex vector space). αu is also written as $u \alpha$ whenever convenient, and $\alpha^{-1} u$ is often written as u/α. The *zero vector* is denoted by 0 and will not be distinguished in symbol from the scalar zero.

Vectors $u_1, \ldots, u_n$ are said to be *linearly independent* if their *linear combination* $\alpha_1 u_1 + \cdots + \alpha_n u_n$ is equal to zero only if $\alpha_1 = \cdots = \alpha_n = 0$; otherwise they are *linearly dependent*. The *dimension* of X, denoted by $\dim \mathsf{X}$, is the largest number of linearly independent vectors that exist in X. If there is no such finite number, we set $\dim \mathsf{X} = \infty$. In the present chapter, all vector spaces are assumed to be finite-dimensional ($0 \leqq \dim \mathsf{X} < \infty$) unless otherwise stated.

[1] See, e. g., GELFAND [[1]], HALMOS [[2]], HOFFMAN and KUNZE [[1]].

A subset M of X is a *linear manifold* or a *subspace* if M is itself a vector space under the same linear operations as in X. The dimension of M does not exceed that of X. For any subset S of X, the set M of all possible linear combinations constructed from the vectors of S is a linear manifold; M is called the linear manifold determined or spanned by S or simply the *(linear) span* of S. According to a basic theorem on vector spaces, the span M of a set of n vectors $u_1, \ldots, u_n$ is at most n-dimensional; it is exactly n-dimensional if and only if $u_1, \ldots, u_n$ are linearly independent.

There is only one 0-dimensional linear manifold of X, which consists of the vector 0 alone and which we shall denote simply by 0.

Example 1.1. The set $\mathsf{X} = \mathrm{C}^N$ of all ordered N-tuples $u = (\xi_j) = (\xi_1, \ldots, \xi_N)$ of complex numbers is an N-dimensional vector space (the complex euclidean space) with the usual definition of the basic operations $\alpha u + \beta v$. Such a vector u is called a *numerical vector*, and is written in the form of a *column vector* (in vertical arrangement of the components ξ_j) or a *row vector* (in horizontal arrangement) according to convenience.

Example 1.2. The set of all complex-valued continuous functions $u: x \to u(x)$ defined on an interval I of a real variable x is an infinite-dimensional vector space, with the obvious definitions of the basic operations $\alpha u + \beta v$. The same is true when, for example, the u are restricted to be functions with continuous derivatives up to a fixed order n. Also the interval I may be replaced by a region[1] in the m-dimensional real euclidean space R^m.

Example 1.3. The set of all solutions of a linear homogeneous differential equation

$$u^{(n)} + a_1(x)\, u^{(n-1)} + \cdots + a_n(x)\, u = 0$$

with continuous coefficients $a_j(x)$ is an n-dimensional vector space, for any solution of this equation is expressed as a linear combination of n *fundamental solutions*, which are linearly independent.

2. Bases

Let X be an N-dimensional vector space and let $x_1, \ldots, x_N$ be a family[2] of N linearly independent vectors. Then their span coincides with X, and each $u \in \mathsf{X}$ can be *expanded* in the form

$$u = \sum_{j=1}^{N} \xi_j x_j \tag{1.1}$$

in a unique way. In this sense the family $\{x_j\}$ is called a *basis*[3] of X, and the scalars ξ_j are called the *coefficients* (or *coordinates*) of u with respect to this basis. The correspondence $u \to (\xi_j)$ is an *isomorphism*

[1] By a region in R^m we mean either an open set in R^m or the union of an open set and all or a part of its boundary.

[2] We use the term "family" to denote a set of elements depending on a parameter.

[3] This is an *ordered basis* (cf. HOFFMAN and KUNZE ⟦1⟧, p. 47).

of X onto C^N (the set of numerical vectors, see Example 1.1) in the sense that it is one to one and preserves the linear operations, that is, $u \to (\xi_j)$ and $v \to (\eta_j)$ imply $\alpha u + \beta v \to (\alpha \xi_j + \beta \eta_j)$.

As is well known, any family $x_1, \ldots, x_p$ of linearly independent vectors can be enlarged to a basis $x_1, \ldots, x_p, x_{p+1}, \ldots, x_N$ by adding suitable vectors $x_{p+1}, \ldots, x_N$.

Example 1.4. In C^N the N vectors $x_j = (\ldots, 0, 1, 0, \ldots)$ with 1 in the j-th place, $j = 1, \ldots, N$, form a basis (the *canonical basis*). The coefficients of $u = (\xi_j)$ with respect to the canonical basis are the ξ_j themselves.

Any two bases $\{x_j\}$ and $\{x'_j\}$ of X are connected by a system of linear relations

$$x_k = \sum_j \gamma_{jk} x'_j, \quad k = 1, \ldots, N. \tag{1.2}$$

The coefficients ξ_j and ξ'_j of one and the same vector u with respect to the bases $\{x_j\}$ and $\{x'_j\}$ respectively are then related to each other by

$$\xi'_j = \sum_k \gamma_{jk} \xi_k, \quad j = 1, \ldots, N. \tag{1.3}$$

The inverse transformations to (1.2) and (1.3) are

$$x'_j = \sum_k \hat{\gamma}_{kj} x_k, \quad \xi_k = \sum_j \hat{\gamma}_{kj} \xi'_j, \tag{1.4}$$

where $(\hat{\gamma}_{jk})$ is the inverse of the matrix (γ_{jk}):

$$\sum_i \hat{\gamma}_{ji} \gamma_{ik} = \sum_i \gamma_{ji} \hat{\gamma}_{ik} = \delta_{jk} = \begin{cases} 1 & (j = k) \\ 0 & (j \neq k), \end{cases} \tag{1.5}$$

$$\det(\gamma_{jk}) \det(\hat{\gamma}_{jk}) = 1. \tag{1.6}$$

Here $\det(\gamma_{jk})$ denotes the determinant of the matrix (γ_{jk}).

The systems of linear equations (1.3) and (1.4) are conveniently expressed by the matrix notation

$$(u)' = (C)(u), \quad (u) = (C)^{-1}(u)', \tag{1.7}$$

where (C) is the matrix (γ_{jk}), $(C)^{-1}$ is its inverse and (u) and $(u)'$ stand for the *column vectors* with components ξ_j and ξ'_j respectively. It should be noticed that (u) or $(u)'$ is conceptually different from the "abstract" vector u which it represents in a particular choice of the basis.

3. Linear manifolds

For any subset S and S' of X, the symbol $\mathsf{S} + \mathsf{S}'$ is used to denote the *(linear) sum* of S and S', that is, the set of all vectors of the form $u + u'$ with $u \in \mathsf{S}$ and $u' \in \mathsf{S}'$[1]. If S consists of a single vector u, $\mathsf{S} + \mathsf{S}'$

[1] $\mathsf{S} + \mathsf{S}'$ should be distinguished from the *union* of S and S', denoted by $\mathsf{S} \cup \mathsf{S}'$. The *intersection* of S and S' is denoted by $\mathsf{S} \cap \mathsf{S}'$.

is simply written $u + \mathsf{S}'$. If M is a linear manifold, $u + \mathsf{M}$ is called the *inhomogeneous linear manifold* (or linear variety) through u parallel to M. The totality of the inhomogeneous linear manifolds $u + \mathsf{M}$ with a fixed M becomes a vector space under the linear operation

$$\alpha(u + \mathsf{M}) + \beta(v + \mathsf{M}) = (\alpha u + \beta v) + \mathsf{M} . \tag{1.8}$$

This vector space is called the *quotient space* of X by M and is denoted by X/M. The elements of X/M are also called the *cosets* of M. The zero vector of X/M is the set M, and we have $u + \mathsf{M} = v + \mathsf{M}$ if and only if $u - v \in \mathsf{M}$. The dimension of X/M is called the *codimension* or *deficiency* of M (with respect to X) and is denoted by $\operatorname{codim} \mathsf{M}$. We have

$$\dim \mathsf{M} + \operatorname{codim} \mathsf{M} = \dim \mathsf{X} . \tag{1.9}$$

If M_1 and M_2 are linear manifolds, $\mathsf{M}_1 + \mathsf{M}_2$ and $\mathsf{M}_1 \cap \mathsf{M}_2$ are again linear manifolds, and

$$\dim(\mathsf{M}_1 + \mathsf{M}_2) + \dim(\mathsf{M}_1 \cap \mathsf{M}_2) = \dim \mathsf{M}_1 + \dim \mathsf{M}_2 . \tag{1.10}$$

The operation $\mathsf{M}_1 + \mathsf{M}_2$ for linear manifolds (or for any subsets of X) is associative in the sense that $(\mathsf{M}_1 + \mathsf{M}_2) + \mathsf{M}_3 = \mathsf{M}_1 + (\mathsf{M}_2 + \mathsf{M}_3)$, which is simply written $\mathsf{M}_1 + \mathsf{M}_2 + \mathsf{M}_3$. Similarly we can define $\mathsf{M}_1 + {} + \mathsf{M}_2 + \cdots + \mathsf{M}_s$ for s linear manifolds M_j.

X is the *direct sum* of the linear manifolds $\mathsf{M}_1, \ldots, \mathsf{M}_s$ if $\mathsf{X} = \mathsf{M}_1 + {} + \cdots + \mathsf{M}_s$ and $\sum u_j = 0$ $(u_j \in \mathsf{M}_j)$ implies that all the $u_j = 0$. Then we write

$$\mathsf{X} = \mathsf{M}_1 \oplus \cdots \oplus \mathsf{M}_s . \tag{1.11}$$

In this case each $u \in \mathsf{X}$ has a unique expression of the form

$$u = \sum_j u_j , \quad u_j \in \mathsf{M}_j , \quad j = 1, \ldots, s . \tag{1.12}$$

Also we have

$$\dim \mathsf{X} = \sum_j \dim \mathsf{M}_j . \tag{1.13}$$

Problem 1.5. If $\mathsf{X} = \mathsf{M}_1 \oplus \mathsf{M}_2$, then $\dim \mathsf{M}_2 = \operatorname{codim} \mathsf{M}_1$.

4. Convergence and norms

Let $\{x_j\}$ be a basis in a finite-dimensional vector space X. Let $\{u_n\}$, $n = 1, 2, \ldots$, be a sequence of vectors of X, with the coefficients ξ_{nj} with respect to the basis $\{x_j\}$. The sequence $\{u_n\}$ is said to *converge* to 0 or have *limit* 0, and we write $u_n \to 0$, $n \to \infty$, or $\lim\limits_{n\to\infty} u_n = 0$, if

$$\lim_{n\to\infty} \xi_{nj} = 0 , \quad j = 1, \ldots, N . \tag{1.14}$$

If $u_n - u \to 0$ for some u, $\{u_n\}$ is said to *converge* to u (or have limit u), in symbol $u_n \to u$ or $\lim u_n = u$. The limit is unique when it exists.

This definition of convergence is independent of the basis $\{x_j\}$ employed. In fact, the formula (1.3) for the coordinate transformation shows that (1.14) implies $\lim \xi'_{nj} = 0$, where the ξ'_{nj} are the coefficients of u_n with respect to a new basis $\{x'_j\}$.

The linear operations in X are *continuous* with respect to this notion of convergence, in the sense that $\alpha_n \to \alpha$, $\beta_n \to \beta$, $u_n \to u$ and $v_n \to v$ imply $\alpha_n u_n + \beta_n u_n \to \alpha u + \beta v$.

For various purposes it is convenient to express the convergence of vectors by means of a *norm*. For example, for a fixed basis $\{x_j\}$ of X, set

$$\|u\| = \max_j |\xi_j| , \tag{1.15}$$

where the ξ_j are the coefficients of u with respect to $\{x_j\}$. Then (1.14) shows that $u_n \to u$ is equivalent to $\|u_n - u\| \to 0$. $\|u\|$ is called the norm of u.

(1.15) is not the only possible definition of a norm. We could as well choose

$$\|u\| = \sum_j |\xi_j| \tag{1.16}$$

or

$$\|u\| = \Big(\sum_j |\xi_j|^2\Big)^{1/2} . \tag{1.17}$$

In each case the following conditions are satisfied:

$$\begin{aligned} &\|u\| \geqq 0 ; \quad \|u\| = 0 \quad \text{if and only if} \quad u = 0 . \\ &\|\alpha u\| = |\alpha| \, \|u\| \quad \text{(homogeneity)} . \\ &\|u + v\| \leqq \|u\| + \|v\| \quad \text{(the triangle inequality)} . \end{aligned} \tag{1.18}$$

Any function $\|u\|$ defined for all $u \in \mathsf{X}$ and satisfying these conditions is called a *norm*. Note that the last inequality of (1.18) implies

$$\big| \|u\| - \|v\| \big| \leqq \|u - v\| \tag{1.19}$$

as is seen by replacing u by $u - v$.

A vector u with $\|u\| = 1$ is said to be *normalized*. For any $u \neq 0$, the vector $u_0 = \|u\|^{-1} u$ is normalized; u_0 is said to result from u by *normalization*.

When a norm $\| \ \|$ is given, the convergence $u_n \to u$ can be defined in a natural way by $\|u_n - u\| \to 0$. This definition of convergence is actually independent of the norm employed and, therefore, coincides with the earlier definition. This follows from the fact that any two norms $\| \ \|$

and $\| \ \|$ $\| \ \|'$ in the same space X are *equivalent* in the sense that

$$\alpha' \|u\| \leqq \|u\|' \leqq \beta' \|u\| , \quad u \in \mathsf{X} , \tag{1.20}$$

where α', β' are positive constants independent of u.

We note incidentally that, for *any* norm $\| \ \|$ and *any* basis $\{x_j\}$, the coefficients ξ_j of a vector u satisfy the inequalities

$$|\xi_j| \leqq \gamma \|u\| , \quad j = 1, \ldots, N , \tag{1.21}$$

$$\|u\| \leqq \gamma' \max_j |\xi_j| , \tag{1.22}$$

where γ, γ' are positive constants depending only on the norm $\| \ \|$ and the basis $\{x_j\}$. These inequalities follow from (1.20) by identifying the norm $\| \ \|'$ with the special one (1.15).

A norm $\|u\|$ is a continuous function of u. This means that $u_n \to u$ implies $\|u_n\| \to \|u\|$, and follows directly from (1.19). It follows from the same inequality that $u_n \to u$ implies that $\{u_n\}$ is a *Cauchy sequence*, that is, the *Cauchy condition*

$$\|u_n - u_m\| \to 0 , \quad m, n \to \infty , \tag{1.23}$$

is satisfied. Conversely, it is easy to see that the Cauchy condition is sufficient for the existence of $\lim u_n$.

The introduction of a norm is not indispensable for the definition of the notion of convergence of vectors, but it is a very convenient means for it. For applications it is important to choose a norm most suitable to the purpose. A vector space in which a norm is defined is called a *normed (vector) space*. Any finite-dimensional vector space can be made into a normed space. The same vector space gives rise to different normed spaces by different choices of the norm. In what follows we shall often regard a given vector space as a normed space by introducing an appropriate norm. The notion of a finite-dimensional normed space considered here is a model for (and a special case of) the notion of a *Banach space* to be introduced in later chapters.

5. Topological notions in a normed space

In this paragraph a brief review will be given on the topological notions associated with a normed space[1]. Since we are here concerned primarily with a finite-dimensional space, there is no essential difference from the case of a real euclidean space. The modification needed in the infinite-dimensional spaces will be indicated later.

[1] We shall need only elementary notions in the topology of metric spaces. As a handy textbook, we refer e. g. to ROYDEN [[1]].

A normed space X is a special case of a *metric space* in which the distance between any two points is defined. In X the distance between two points (vectors) u, v is defined by $\|u - v\|$. An (open) *ball* of X is the set of points $u \in \mathsf{X}$ such that $\|u - u_0\| < r$, where u_0 is the *center* and $r > 0$ is the *radius* of the ball. The set of u with $\|u - u_0\| \leqq r$ is a *closed ball*. We speak of the *unit ball* when $u_0 = 0$ and $r = 1$. Given a $u \in \mathsf{X}$, any subset of X containing a ball with center u is called a *neighborhood* of u. A subset of X is said to be *bounded* if it is contained in a ball. X itself is not bounded unless $\dim \mathsf{X} = 0$.

For any subset S of X, u is an *interior point* of S if S is a neighborhood of u. u is an *exterior point* of S if u is an interior point of the complement S' of S (with respect to X). u is a boundary point of S if it is neither an interior nor an exterior point of S. The set $\partial \mathsf{S}$ of all boundary points of S is the *boundary* of S. The union $\bar{\mathsf{S}}$ of S and its boundary is the *closure* of S. S is *open* if it consists only of interior points. S is *closed* if S' is open, or, equivalently, if $\mathsf{S} = \bar{\mathsf{S}}$. The closure of any subset S is closed: $\bar{\bar{\mathsf{S}}} = \bar{\mathsf{S}}$. Every linear manifold of X is closed (X being finite-dimensional).

These notions can also be defined by using convergent sequences. For example, $\bar{\mathsf{S}}$ is the set of all $u \in \mathsf{X}$ such that there is a sequence $u_n \in \mathsf{S}$ with $u_n \to u$. S is closed if and only if $u_n \in \mathsf{S}$ and $u_n \to u$ imply $u \in \mathsf{S}$.

We denote by $\operatorname{dist}(u, \mathsf{S})$ the distance of u from a subset S:

$$\operatorname{dist}(u, \mathsf{S}) = \inf_{v \in \mathsf{S}} \|u - v\| . \tag{1.24}$$

If S is closed and $u \notin \mathsf{S}$, then $\operatorname{dist}(u, \mathsf{S}) > 0$.

An important property of a finite-dimensional normed space X is that the theorem of Bolzano-Weierstrass holds true. From each bounded sequence $\{u_n\}$ of vectors of X, it is possible to extract a subsequence $\{v_n\}$ that converges to some $v \in \mathsf{X}$. This property is expressed by saying that X is *locally compact*[1]. A subset $\mathsf{S} \subset \mathsf{X}$ is *compact* if any sequence of elements of S has a subsequence converging to an element of S.

6. Infinite series of vectors

The convergence of an infinite series

$$\sum_{n=1}^{\infty} u_n \tag{1.25}$$

of vectors $u_n \in \mathsf{X}$ is defined as in the case of numerical series. (1.25) is said to converge to v (or have the sum v) if the sequence $\{v_n\}$ consisting of the partial sums $v_n = \sum_{k=1}^{n} u_k$ converges (to v). The sum v is usually denoted by the same expression (1.25) as the series itself.

[1] The proof of (1.20) depends essentially on the local compactness of X.

A sufficient condition for the convergence of (1.25) is

$$\sum_n \|u_n\| < \infty . \tag{1.26}$$

If this is true for some norm, it is true for any norm in virtue of (1.20). In this case the series (1.25) is said to *converge absolutely*. We have

$$\|\sum_n u_n\| \leqq \sum_n \|u_n\| . \tag{1.27}$$

Problem 1.6. If u_n and v have respectively the coefficients ξ_{nj} and η_j with respect to a basis $\{x_j\}$, (1.25) converges to v if and only if $\sum_n \xi_{nj} = \eta_j$, $j = 1, \ldots, N$. (1.25) converges absolutely if and only if the N numerical series $\sum_n \xi_{nj}$, $j = 1, \ldots, N$, converge absolutely.

In an absolutely convergent series of vectors, the order of the terms may be changed arbitrarily without affecting the sum. This is obvious if we consider the coefficients with respect to a basis (see Problem 1.6). For later reference, however, we shall sketch a more direct proof without using the coefficients. Let $\sum u_n'$ be a series obtained from (1.25) by changing the order of terms. It is obvious that $\sum \|u_n'\| = \sum \|u_n\| < \infty$. For any $\varepsilon > 0$, there is an integer m such that $\sum_{n=m+1}^{\infty} \|u_n\| < \varepsilon$. Let p be so large that $u_1, \ldots, u_m$ are contained in $u_1', \ldots, u_p'$. For any $n > m$ and $q > p$, we have then $\left\| \sum_{j=1}^{q} u_j' - \sum_{k=1}^{n} u_k \right\| \leqq \sum_{k=m+1}^{\infty} \|u_k\| < \varepsilon$, and going to the limit $n \to \infty$ we obtain $\left\| \sum_{j=1}^{q} u_j' - \sum_{k=1}^{\infty} u_k \right\| \leqq \varepsilon$ for $q > p$. This proves that $\sum u_n' = \sum u_n$.

This is an example showing how various results on numerical series can be taken over to series of vectors. In a similar way it can be proved, for example, that an absolutely convergent double series of vectors may be summed in an arbitrary order, by rows or by columns or by transformation into a simple series.

7. Vector-valued functions

Instead of a sequence $\{u_n\}$ of vectors, which may be regarded as a function from the set $\{n\}$ of integers into X, we may consider a function $u_t = u(t)$ defined for a real or complex variable t and taking values in X. The relation $\lim_{t \to a} u(t) = v$ is defined by $\|u(t) - v\| \to 0$ for $t \to a$ (with the usual understanding that $t \neq a$) with the aid of *any* norm. $u(t)$ is *continuous* at $t = a$ if $\lim_{t \to a} u(t) = u(a)$, and $u(t)$ is continuous in a region E of t if it is continuous at every point of E.

The derivative of $u(t)$ is given by

$$u'(t) = \frac{du(t)}{dt} = \lim_{h \to 0} h^{-1}(u(t+h) - u(t)) \tag{1.28}$$

whenever this limit exists. The formulas

$$\begin{aligned} &\frac{d}{dt}(u(t) + v(t)) = u'(t) + v'(t)\,, \\ &\frac{d}{dt}\phi(t)\, u(t) = \phi(t)\, u'(t) + \phi'(t)\, u(t) \end{aligned} \tag{1.29}$$

are valid exactly as for numerical functions, where $\phi(t)$ denotes a complex-valued function.

The integral of a vector-valued function $u(t)$ can also be defined as for numerical functions. For example, suppose that $u(t)$ is a continuous function of a real variable t, $a \leqq t \leqq b$. The Riemann integral $\int_a^b u(t)\, dt$ is defined as an appropriate limit of the sums $\sum (t_j - t_{j-1})\, u(t_j)$ constructed for the partitions $a = t_0 < t_1 < \cdots < t_n = b$ of the interval $[a, b]$. Similarly an integral $\int_C u(t)\, dt$ can be defined for a continuous function $u(t)$ of a complex variable t and for a rectifiable curve C. The proof of the existence of such an integral is quite the same as for numerical functions; in most cases it is sufficient to replace the absolute value of a complex number by the norm of a vector. For these integrals we have the formulas

$$\begin{aligned} &\int (\alpha\, u(t) + \beta\, v(t))\, dt = \alpha \int u(t)\, dt + \beta \int v(t)\, dt\,, \\ &\left\| \int u(t)\, dt \right\| \leqq \int \|u(t)\|\, |dt|\,. \end{aligned} \tag{1.30}$$

There is no difficulty in extending these definitions to improper integrals. We shall make free use of the formulas of differential and integral calculus for vector-valued functions without any further comments.

Although there is no difference in the formal definition of the derivative of a vector-valued function $u(t)$ whether the variable t is real or complex, there is an essential difference between these two cases just as with numerical functions. When $u(t)$ is defined and differentiable everywhere in a domain D of the complex plane, $u(t)$ is said to be *regular (analytic)* or *holomorphic* in D. Most of the results of complex function theory are applicable to such vector-valued, holomorphic functions[1].

---

[1] Throughout this book we shall make much use of complex function theory, but it will be limited to elementary results given in standard textbooks such as KNOPP [[1, 2]]. Actually we shall apply these results to vector- or operator-valued functions as well as to complex-valued functions, but such a generalization usually offers no difficulty and we shall make it without particular comments. For the theorems used we shall refer to Knopp whenever necessary.

Thus we have Cauchy's integral theorem, Taylor's and Laurent's expansions, Liouville's theorem, and so on. For example, if $t = 0$ is an isolated singularity of a holomorphic function $u(t)$, we have

$$(1.31) \qquad u(t) = \sum_{n=-\infty}^{+\infty} t^n a_n , \quad a_n = \frac{1}{2\pi i} \int_C t^{-n-1} u(t)\, dt ,$$

where C is a closed curve, say a circle, enclosing $t = 0$ in the positive direction. $t = 0$ is a regular point (removable singularity) if $a_n = 0$ for $n < 0$, a *pole* of order $k > 0$ if $a_{-k} \neq 0$ whereas $a_n = 0$ for $n < -k$, and an *essential singularity* otherwise[1].

Problem 1.7. If $t = 0$ is a pole of order k, then $\|u(t)\| = O(|t|^{-k})$ for $t \to 0$.

Problem 1.8. Let $\xi_j(t)$ be the coefficients of $u(t)$ with respect to a basis of X. $u(t)$ is continuous (differentiable) if and only if all the $\xi_j(t)$ are continuous (differentiable). $u'(t)$ has the coefficients $\xi_j'(t)$ for the same basis. Similarly, $\int u(t)\, dt$ has the coefficients $\int \xi_j(t)\, dt$.

§ 2. Linear forms and the adjoint space

1. Linear forms

Let X be a vector space. A complex-valued function $f[u]$ defined for $u \in \mathsf{X}$ is called a *linear form* or a *linear functional* if

$$(2.1) \qquad f[\alpha u + \beta v] = \alpha f[u] + \beta f[v]$$

for all u, v of X and all scalars α, β.

Example 2.1. If $\mathsf{X} = \mathsf{C}^N$ (the space of N-dimensional numerical vectors), a linear form on X can be expressed in the form

$$(2.2) \qquad f[u] = \sum_{j=1}^{N} \alpha_j \xi_j \quad \text{for} \quad u = (\xi_j) .$$

It is usual to represent f as a *row vector* with the components α_j, when u is represented as a *column vector* with the components ξ_j. (2.2) is the matrix product of these two vectors.

Example 2.2. Let X be the space of continuous functions $u = u(x)$ considered in Example 1.2. The following are examples of linear forms on X:

$$(2.3) \qquad f[u] = u(x_0) , \quad x_0 \text{ being fixed.}$$

$$(2.4) \qquad f[u] = \int_a^b \phi(x)\, u(x)\, dx , \quad \phi(x) \text{ being a given function.}$$

Let $\{x_j\}$ be a basis of X ($\dim \mathsf{X} = N < \infty$). If $u = \sum \xi_j x_j$ is the expansion of u, we have by (2.1)

$$(2.5) \qquad f[u] = \sum \alpha_j \xi_j$$

where $\alpha_j = f[x_j]$. Each linear form is therefore represented by a numerical vector (α_j) with respect to the basis and, conversely, each numerical

[1] See KNOPP [1], p. 117.

vector (α_j) determines a linear form f by (2.5). (2.5) corresponds exactly to (2.2) for a linear form on C^N.

The same linear form f is represented by a different numerical vector (α_j') for a different basis $\{x_j'\}$. If the new basis is connected with the old one through the transformation (1.2) or (1.4), the relation between these representations is given by

$$\begin{aligned} \alpha_j' &= f[x_j'] = \sum_k \hat{\gamma}_{kj} f[x_k] = \sum_k \hat{\gamma}_{kj} \alpha_k , \\ \alpha_k &= \sum_j \gamma_{jk} \alpha_j' . \end{aligned} \tag{2.6}$$

In the matrix notation, these may be written

$$(f)' = (f) (C)^{-1} , \quad (f) = (f)' (C) , \tag{2.7}$$

where (C) is the matrix (γ_{jk}) [see (1.7)] and where (f) and $(f)'$ stand for the *row vectors* with components (α_j) and (α_j') respectively.

2. The adjoint space

A complex-valued function $f[u]$ defined on X is called a *semilinear* (or *conjugate-linear* or *anti-linear*) form if

$$f[\alpha u + \beta v] = \bar{\alpha} f[u] + \bar{\beta} f[v] , \tag{2.8}$$

where $\bar{\alpha}$ denotes the complex conjugate of α. It is obvious that $f[u]$ is a semilinear form if and only if $\overline{f[u]}$ is a linear form. For the sake of a certain formal convenience, we shall hereafter be concerned with semilinear rather than with linear forms.

Example 2.3. A semilinear form on C^N is given by (2.2) with the ξ_j on the right replaced by the $\bar{\xi}_j$, where $u = (\xi_j)$.

Example 2.4. Let X be as in Example 2.2. The following are examples of semilinear forms on X:

$$f[u] = \overline{u(x_0)} , \tag{2.9}$$

$$f[u] = \int_a^b \phi(x) \overline{u(x)} \, dx . \tag{2.10}$$

The linear combination $\alpha f + \beta g$ of two semilinear forms f, g defined by

$$(\alpha f + \beta g) [u] = \alpha f[u] + \beta g[u] \tag{2.11}$$

is obviously a semilinear form. Thus the set of all semilinear forms on X becomes a vector space, called the *adjoint* (or *conjugate*) *space* of X and denoted by X^*. The zero vector of X^*, which is again denoted by 0, is the zero form that sends every vector u of X into the complex number zero.

It is convenient to treat X^* on the same level as X. To this end we write

$$f[u] = (f, u) \tag{2.12}$$

and call (f, u) the *scalar product* of $f \in X^*$ and $u \in X$. It follows from the definition that (f, u) is linear in f and semilinear in u:

$$\begin{aligned} (\alpha f + \beta g, u) &= \alpha (f, u) + \beta (g, u) , \\ (f, \alpha u + \beta v) &= \bar{\alpha} (f, u) + \bar{\beta} (f, v) . \end{aligned} \tag{2.13}$$

Example 2.5. For $X = C^N$, X^* may be regarded as the set of all row vectors $f = (\alpha_j)$ whereas X is the set of all column vectors $u = (\xi_j)$. Their scalar product is given by

$$(f, u) = \sum \alpha_j \bar{\xi}_j . \tag{2.14}$$

Remark 2.6. In the algebraic theory of vector spaces, the *dual space* of a vector space X is defined to be the set of all *linear* forms on X. Our definition of the adjoint space is chosen in such a way that the adjoint space of a unitary space (see § 6) X can be identified with X itself[1].

3. The adjoint basis

Let $\{x_j\}$ be a basis of X. As in the case of linear forms, for each numerical vector (α_k) there is an $f \in X^*$ such that $(f, x_k) = \alpha_k$. In particular, it follows that for each j, there exists a unique $e_j \in X^*$ such that

$$(e_j, x_k) = \delta_{jk} , \quad j, k = 1, \ldots, N . \tag{2.15}$$

It is easy to see that the e_j are linearly independent. Each $f \in X^*$ can be expressed in a unique way as a linear combination of the e_j, according to

$$f = \sum_j \alpha_j e_j \quad \text{where} \quad \alpha_j = (f, x_j) . \tag{2.16}$$

In fact, the difference of the two members of (2.16) has scalar product zero with all the x_k and therefore with all $u \in X$; thus it must be equal to the zero form.

Thus the N vectors e_j form a basis of X^*, called the basis *adjoint* to the basis $\{x_j\}$ of X. Since the basis $\{e_j\}$ consists of N elements, we have

$$\dim X^* = \dim X = N . \tag{2.17}$$

For each $u \in X$ we have

$$u = \sum_j \xi_j x_j \quad \text{where} \quad \xi_j = \overline{(e_j, u)} . \tag{2.18}$$

[1] See e. g. HALMOS [2]. Sometimes one defines X^* as the set of all linear forms f on X but defines αf by $(\alpha f) [u] = \bar{\alpha} f [u]$, so that $f[u]$ is linear in u and semilinear in f (see e. g. LORCH [1]). Our definition of X^* is the same as in RIESZ and SZ.-NAGY [1] in this respect.

It follows from (2.16) and (2.18) that

$$(f, u) = \sum \alpha_j \bar{\xi}_j = \sum (f, x_j)(e_j, u) . \tag{2.19}$$

Let $\{x_j\}$ and $\{x'\}$ be two bases of X related to each other by (1.2). Then the corresponding adjoint bases $\{e_j\}$ and $\{e'_j\}$ of X^* are related to each other by the formulas

$$e'_j = \sum_k \bar{\gamma}_{jk} e_k , \quad e_k = \sum_j \bar{\hat{\gamma}}_{kj} e'_j . \tag{2.20}$$

Furthermore we have

$$\bar{\gamma}_{jk} = (e'_j, x_k) , \quad \bar{\hat{\gamma}}_{kj} = (e_k, x'_j) . \tag{2.21}$$

4. The adjoint space of a normed space

Since X^* is an N-dimensional vector space with X, the notion of convergence of a sequence of vectors of X^* is defined as in § 1.4. For the same reason a norm could be introduced into X^*. Usually the norm in X^* is not defined independently but is correlated with the norm of X.

When a norm $\|u\|$ in X is given so that X is a normed space, X^* is by definition a normed space with the norm $\|f\|$ defined by[1]

$$\|f\| = \sup_{0 \neq u \in \mathsf{X}} \frac{|(f, u)|}{\|u\|} = \sup_{\|u\| = 1} |(f, u)| . \tag{2.22}$$

That $\|f\|$ is finite follows from the fact that the continuous function $|(f, u)|$ of u attains a maximum for $\|u\| = 1$ (because X is locally compact). It is easily verified that the norm $\|f\|$ thus defined satisfies the conditions (1.18) of a norm. There is no fear of confusion in using the same symbol $\| \ \|$ for the two norms.

Example 2.7. Suppose that the norm in X is given by (1.15) for a fixed basis $\{x_j\}$. If $\{e_j\}$ is the adjoint basis in X^*, we have $|(f, u)| \leqq (\sum |\alpha_j|) \|u\|$ by (2.19). But the equality holds if u is such that $|\xi_1| = |\xi_2| = \cdots = |\xi_N|$ and all $\alpha_j \bar{\xi}_j$ are real and nonnegative. This shows that

$$\|f\| = \sum |\alpha_j| . \tag{2.23}$$

Similarly it can be shown that, when the norm in X is given by (1.16), the norm in X^* is given by

$$\|f\| = \max |\alpha_j| . \tag{2.24}$$

Thus we may say that the norms (1.15) and (1.16) are adjoint to each other.

(2.22) shows that

$$|(f, u)| \leq \|f\| \, \|u\| , \quad f \in \mathsf{X}^* , \quad u \in \mathsf{X} . \tag{2.25}$$

This is called the *Schwarz inequality* in the generalized sense. As we have deduced it, it is simply the definition of $\|f\|$ and has an essential meaning only when we give $\|f\|$ some independent characterization (as, for example, in the case of a unitary space; see § 6).

---

[1] Here we assume $\dim \mathsf{X} > 0$; the case $\dim \mathsf{X} = 0$ is trivial.

(2.25) implies that $\|u\| \geqq |(f, u)|/\|f\|$. Actually the following stronger relation is true[1]:

$$\|u\| = \sup_{0 \neq f \in X^*} \frac{|(f, u)|}{\|f\|} = \sup_{\|f\|=1} |(f, u)| . \tag{2.26}$$

This follows from the fact that, for any $u_0 \in X$, there is an $f \in X^*$ such that

$$(f, u_0) = \|u_0\| , \quad \|f\| = 1 . \tag{2.27}$$

The proof of (2.27) requires a deeper knowledge of the nature of a norm and will be given in the following paragraph.

Problem 2.8. $(f, u) = 0$ for all $u \in X$ implies $f = 0$. $(f, u) = 0$ for all $f \in X^*$ implies $u = 0$.

A simple consequence of the Schwarz inequality is the fact that *the scalar product* (f, u) *is a continuous function of* f *and* u. In fact[2],

$$\begin{aligned} |(f', u') - (f, u)| &= |(f' - f, u) + (f, u' - u) + (f' - f, u' - u)| \\ &\leqq \|f' - f\| \, \|u\| + \|f\| \, \|u' - u\| + \|f' - f\| \, \|u' - u\| . \end{aligned} \tag{2.28}$$

In particular, $u_n \to u$ implies $(f, u_n) \to (f, u)$ for every $f \in X^*$ and $f_n \to f$ implies $(f_n, u) \to (f, u)$ for every $u \in X$. Similarly, the convergence of a series $\sum u_n = u$ implies the convergence $\sum (f, u_n) = (f, u)$ for every $f \in X^*$ (that is, term by term multiplication is permitted for the scalar product). Conversely, $(f, u_n) \to (f, u)$ for all $f \in X^*$ implies $u_n \to u$; this can be seen by expanding u_n and u by a fixed basis of X.

Problem 2.8a. If $u_1, \ldots, u_r$ are linearly independent, they remain linearly independent when changed slightly. [hint: extend $u_1, \ldots, u_r$ to a basis $\{u_j\}$ of X. Let $\{e_k\}$ be the adjoint basis of X^*. Then $\det((u_j, e_k)) = 1$. If $u_1, \ldots, u_r$ are changed slightly, with the other u_j and e_k fixed, the determinant is still different from zero.]

5. The convexity of balls

Let S be an open ball of X. S is a *convex* set: for any two points (vectors) u, v of S, the segment joining u and v belongs to S. In other words,

$$\lambda u + (1 - \lambda) v \in S \quad \text{if} \quad u, v \in S \quad \text{and} \quad 0 \leqq \lambda \leqq 1 . \tag{2.29}$$

In fact, denoting by u_0 the center and by r the radius of S, we have $\|\lambda u + (1 - \lambda) v - u_0\| = \|\lambda (u - u_0) + (1 - \lambda) (v - u_0)\| < \lambda r + (1 - \lambda) r = r$, which proves the assertion. In what follows we assume S to be the unit ball ($u_0 = 0$, $r = 1$).

Since X is isomorphic with the N-dimensional complex euclidean space C^N, X is isomorphic with the $2N$-dimensional real euclidean

[1] Again $\dim X^* = \dim X > 0$ is assumed.

[2] The continuity of (f, u) follows immediately from (2.19). But the proof in (2.28) has the advantage that it is valid in the ∞-dimensional case.

space R^{2N} as a *real vector space* (that is, when only real numbers are regarded as scalars). Thus S may be regarded as a convex set in R^{2N}. It follows from a well-known theorem[1] on convex sets in R^n that, for each vector u_0 lying on the boundary of S (that is, $\|u_0\| = 1$), there is a *support hyperplane* of S through u_0. This implies that there exists a *real-linear* form $g[u]$ on X such that

$$g[u_0] = 1 \quad \text{whereas} \quad g[u] < 1 \quad \text{for} \quad u \in \mathsf{S}\,. \tag{2.30}$$

That g is real-linear means that $g[u]$ is real-valued and $g[\alpha u + \beta v] = \alpha g[u] + \beta g[v]$ for all real numbers α, β and $u, v \in \mathsf{X}$.

g is neither a linear nor a semilinear form on the *complex* vector space X. But there is an $f \in \mathsf{X}^*$ related to g according to[2]

$$(f, u) = f[u] = g[u] + i\, g[i\, u]\,. \tag{2.31}$$

To see that this f is in fact a semilinear form on X, it suffices to verify that $f[(\alpha + i\,\beta)\, u] = (\alpha - i\,\beta)\, f[u]$ for real α, β, for it is obvious that $f[u + v] = f[u] + f[v]$. This is seen as follows:

$$\begin{aligned} f[(\alpha + i\,\beta)\, u] &= g[\alpha\, u + i\,\beta\, u] + i\, g[i\,\alpha\, u - \beta\, u] \\ &= \alpha\, g[u] + \beta\, g[i\, u] + i\,\alpha\, g[i\, u] - i\,\beta\, g[u] \\ &= (\alpha - i\,\beta)\,(g[u] + i\, g[i\, u]) = (\alpha - i\,\beta)\, f[u]\,. \end{aligned}$$

Now this f has the following properties:

$$(f, u_0) = 1\,, \quad \|f\| = 1\,. \tag{2.32}$$

To see this, set $(f, u) = R e^{i\theta}$, θ real and $R \geqq 0$. It follows from what was just proved that $(f, e^{i\theta} u) = e^{-i\theta}(f, u) = R$ and hence that $|(f, u)| = R = \mathrm{Re}\,(f, e^{i\theta} u) = g[e^{i\theta} u] < 1$ if $\|e^{i\theta} u\| = \|u\| < 1$. This shows that $\|f\| \leqq 1$. In particular we have $|(f, u_0)| \leqq 1$. But since $\mathrm{Re}\,(f, u_0) = g[u_0] = 1$, we must have $(f, u_0) = 1$. This implies also $\|f\| = 1$.

Note that (2.32) is equivalent to (2.27) in virtue of the homogeneity of the norm.

6. The second adjoint space

The adjoint space X^{**} to X^* is the aggregate of semilinear forms on X^*. An example of such a semilinear form F is given by $F[f] = \overline{(f, u)}$ where $u \in \mathsf{X}$ is fixed. With each $u \in \mathsf{X}$ is thus associated an element F of X^{**}. This correspondence of X with X^{**} is *linear* in the sense that $\alpha\, u + \beta\, v$ corresponds to $\alpha F + \beta G$ when u, v correspond to F, G, respectively. The fact that $\dim \mathsf{X}^{**} = \dim \mathsf{X}^* = \dim \mathsf{X}$ shows that the whole space X^{**} is exhausted in this way; in other words, to each

---

[1] See, e. g., EGGLESTON [[1]].

[2] i is the imaginary unit.

$F \in \mathsf{X}^{**}$ corresponds a $u \in \mathsf{X}$. Furthermore when X and therefore X^*, X^{**} are normed spaces, the norm in X^{**} is identical with the norm in X: $\|F\| = \|u\|$, as is seen from (2.26). In this way we see that X^{**} can be *identified* with X, not only as a vector space but as a normed space. In this sense we may write $F[f]$ as $u[f] = (u, f)$, so that

$$(u, f) = \overline{(f, u)} \,. \tag{2.33}$$

It should be noted that these results are essentially based on the assumption that $\dim \mathsf{X}$ is finite.

Problem 2.9. If $\{e_j\}$ is the basis of X^* adjoint to the basis $\{x_j\}$ of X, $\{x_j\}$ is the basis of $\mathsf{X}^{**} = \mathsf{X}$ adjoint to $\{e_j\}$.

We write $f \perp u$ or $u \perp f$ when $(f, u) = 0$. When $f \perp u$ for all u of a subset S of X, we write $f \perp \mathsf{S}$. Similarly we introduce the notation $u \perp \mathsf{S}'$ for $u \in \mathsf{X}$ and $\mathsf{S}' \subset \mathsf{X}^*$. The set of all $f \in \mathsf{X}^*$ such that $f \perp \mathsf{S}$ is called the *annihilator* of S and is denoted by $\mathsf{S}^\perp$. Similarly the annihilator $\mathsf{S}'^\perp$ of a subset S' of X^* is the set of all $u \in \mathsf{X}$ such that $u \perp \mathsf{S}'$.

For any $\mathsf{S} \subset \mathsf{X}$, $\mathsf{S}^\perp$ is a linear manifold. *The annihilator* $\mathsf{S}^{\perp\perp}$ *of* $\mathsf{S}^\perp$ *is identical with the linear span* M *of* S. In particular we have $\mathsf{M}^{\perp\perp} = \mathsf{M}$ for any linear manifold M of X.

Problem 2.10. $\operatorname{codim} \mathsf{M} = \dim \mathsf{M}^\perp$.

§ 3. Linear operators

1. Definitions. Matrix representations

Let X, Y be two vector spaces. A function T that sends every vector u of X into a vector $v = Tu$ of Y is called a *linear transformation* or a *linear operator* on X to Y if T preserves linear relations, that is, if

$$T(\alpha_1 u_1 + \alpha_2 u_2) = \alpha_1 T u_1 + \alpha_2 T u_2 \tag{3.1}$$

for all u_1, u_2 of X and all scalars α_1, α_2. X is the *domain space* and Y is the *range space* of T. If $\mathsf{Y} = \mathsf{X}$ we say simply that T is a linear operator *in* X. In this book an operator means a linear operator unless otherwise stated.

For any subset S of X, the set of all vectors of the form Tu with $u \in \mathsf{S}$ is called the *image* under T of S and is denoted by $T\mathsf{S}$; it is a subset of Y. If M is a linear manifold of X, $T\mathsf{M}$ is a linear manifold of Y. In particular, the linear manifold $T\mathsf{X}$ of Y is called the *range* of T and is denoted by $\mathsf{R}(T)$. The dimension of $\mathsf{R}(T)$ is called the *rank* of T; we denote it by $\operatorname{rank} T$. The deficiency (codimension) of $\mathsf{R}(T)$ with respect to Y is called the *deficiency* of T and is denoted by $\operatorname{def} T$. Thus

$$\operatorname{rank} T + \operatorname{def} T = \dim \mathsf{Y} \,. \tag{3.2}$$

For any subset S' of Y, the set of all vectors $u \in X$ such that $Tu \in S'$ is called the *inverse image* of S' and is denoted by $T^{-1} S'$. The inverse image of $0 \subset Y$ is a linear manifold of X; it is called the *kernel* or *null space* of T and is denoted by $N(T)$. The dimension of $N(T)$ is called the *nullity* of T, which we shall denote by $\operatorname{nul} T$. We have

$$\operatorname{rank} T + \operatorname{nul} T = \dim X . \tag{3.3}$$

To see this it suffices to note that T maps the quotient space $X/N(T)$ (which has dimension $\dim X - \operatorname{nul} T$) onto $R(T)$ in a one-to-one fashion.

If both $\operatorname{nul} T$ and $\operatorname{def} T$ are zero, then T maps X onto Y one to one. In this case the *inverse operator* T^{-1} is defined; T^{-1} is the operator on Y to X that sends Tu into u. Obviously we have $(T^{-1})^{-1} = T$. T is said to be *nonsingular* if T^{-1} exists and *singular* otherwise. For T to be nonsingular it is necessary that $\dim X = \dim Y$. If $\dim X = \dim Y$, each of $\operatorname{nul} T = 0$ and $\operatorname{def} T = 0$ implies the other and therefore the nonsingularity of T.

Let $\{x_k\}$ be a basis of X. Each $u \in X$ has the expansion (1.1), so that

$$Tu = \sum_{k=1}^{N} \xi_k \, T x_k , \quad N = \dim X . \tag{3.4}$$

Thus an operator T on X to Y is determined by giving the values of Tx_k, $k = 1, \ldots, N$. Furthermore, these values can be prescribed arbitrarily in Y; then it suffices to define T by (3.4) to make T linear.

If $\{y_j\}$ is a basis of Y, each Tx_k has the expansion

$$T x_k = \sum_{j=1}^{M} \tau_{jk} \, y_j , \quad M = \dim Y . \tag{3.5}$$

Substituting (3.5) into (3.4), we see that the coefficients η_j of $v = Tu$ with respect to the basis $\{y_j\}$ are given by

$$\eta_j = \sum_k \tau_{jk} \, \xi_k , \quad j = 1, \ldots, M . \tag{3.6}$$

In this way an operator T on X to Y is represented by an $M \times N$ matrix (τ_{jk}) with respect to the bases $\{x_k\}$, $\{y_j\}$ of X, Y, respectively. Conversely, to each $M \times N$ matrix (τ_{jk}) there is an operator T on X to Y represented by it with respect to the given bases.

Let (τ'_{jk}) be the matrix representing the same operator T with respect to a new pair of bases $\{x'_k\}$, $\{y'_j\}$. The relationship between the matrices (τ'_{jk}) and (τ_{jk}) is obtained by combining (3.5) and a similar expression for Tx'_k in terms of $\{y'_j\}$ with the formulas (1.2), (1.4) of the coordinate transformation and the corresponding formulas in Y. The result is

$$\tau'_{jk} = \sum_{i,h} \gamma'_{ji} \, \tau_{ih} \, \hat{\gamma}_{hk} . \tag{3.7}$$

Thus the matrix (τ'_{jk}) is the product of three matrices (γ'_{jk}), (τ_{jk}) and $(\hat{\gamma}_{jk})$. [(γ'_{jk}) is the matrix transforming from (γ_k) to (γ'_j).]

If T is an operator on X to itself, it is usual to set $y_j = x_j$ and $y'_j = x'_j$; we have then

$$(3.8) \qquad (\tau'_{jk}) = (\gamma_{jk})\,(\tau_{jk})\,(\hat{\gamma}_{jk})\,.$$

It follows by (1.6) that

$$(3.9) \qquad \det(\tau'_{jk}) = \det(\tau_{jk})\,.$$

Thus $\det(\tau_{jk})$ is determined by the operator T itself and does not depend on the basis employed. It is called the *determinant* of T and is denoted by $\det T$. Similarly, the trace $\sum \tau_{jj}$ of the matrix (τ_{jk}) does not depend on the basis; it is called the *trace* of T and is denoted by $\operatorname{tr} T$.

Problem 3.1. If $\{f_j\}$ is the basis of Y^* adjoint to $\{y_j\}$, then

$$(3.10) \qquad \tau_{jk} = (T x_k, f_j)\,.$$

Problem 3.2. Let $\{x_j\}$ and $\{e_j\}$ be the bases of X and X^*, respectively, which are adjoint to each other. If T is an operator on X to itself, we have

$$(3.11) \qquad \operatorname{tr} T = \sum_j (T x_j, e_j)\,.$$

2. Linear operations on operators

If T and S are two linear operators on X to Y, their linear combination $\alpha S + \beta T$ is defined by

$$(3.12) \qquad (\alpha S + \beta T)\,u = \alpha(S u) + \beta(T u)$$

for all $u \in \mathsf{X}$, and is again a linear operator on X to Y. Let us denote by $\mathscr{B}(\mathsf{X}, \mathsf{Y})$ the set of all operators on X to Y; $\mathscr{B}(\mathsf{X}, \mathsf{Y})$ is a vector space with the linear operations defined as above. The zero vector of this vector space is the zero operator 0 defined by $0\,u = 0$ for all $u \in \mathsf{X}$.

Problem 3.3. $\operatorname{rank}(S + T) \leq \operatorname{rank} S + \operatorname{rank} T$.

The dimension of the vector space $\mathscr{B}(\mathsf{X}, \mathsf{Y})$ is equal to $N M$, where $N = \dim \mathsf{X}$ and $M = \dim \mathsf{Y}$. To see this, let $\{x_k\}$ and $\{y_j\}$ be bases of X and Y, respectively, and let P_{jk} be the operator on X to Y such that

$$(3.13) \qquad P_{jk}\, x_h = \delta_{kh}\, y_j\,, \quad k, h = 1, \ldots, N; \quad j = 1, \ldots, M\,.$$

These $M N$ operators P_{jk} are linearly independent elements of $\mathscr{B}(\mathsf{X}, \mathsf{Y})$, and we have from (3.5)

$$(3.14) \qquad T = \sum \tau_{jk}\, P_{jk}\,.$$

Thus $\{P_{jk}\}$ is a basis of $\mathscr{B}(\mathsf{X}, \mathsf{Y})$, which proves the assertion. $\{P_{jk}\}$ will be called the basis of $\mathscr{B}(\mathsf{X}, \mathsf{Y})$ associated with the bases $\{x_k\}$ and $\{y_j\}$ of X and Y, respectively. (3.14) shows that the matrix elements τ_{jk} are the coefficients of the "vector" T with respect to the basis $\{P_{jk}\}$, and (3.7) or (3.8) is the formula for coordinate transformation in $\mathscr{B}(\mathsf{X}, \mathsf{Y})$.

The product TS of two linear operators T, S is defined by

$$(TS)\,u = T(Su) \tag{3.15}$$

for all $u \in \mathsf{X}$, where X is the domain space of S, provided the domain space of T is identical with the range space Y of S. The following relations hold for these operations on linear operators:

$$\begin{aligned} &(TS)\,R = T(SR), \text{ which is denoted by } TSR\,, \\ &(\alpha T)\,S = T(\alpha S) = \alpha(TS), \text{ denoted by } \alpha TS\,, \\ &(T_1 + T_2)\,S = T_1 S + T_2 S\,, \\ &T(S_1 + S_2) = TS_1 + TS_2\,. \end{aligned} \tag{3.16}$$

Problem 3.4. $\operatorname{rank}(TS) \leqq \max(\operatorname{rank} T, \operatorname{rank} S)$.

Problem 3.5. If S, T have the matrices (σ_{jk}), (τ_{jk}) with respect to some fixed bases, $S + T$ and TS have the matrices $(\sigma_{jk}) + (\tau_{jk})$, $(\tau_{jk})\,(\sigma_{jk})$ respectively (whenever meaningful). If T^{-1} exists, its matrix is the inverse matrix of (τ_{jk}).

3. The algebra of linear operators

If S and T are operators on X to itself, their product TS is defined and is again an operator on X to itself. Thus the set $\mathscr{B}(\mathsf{X}) = \mathscr{B}(\mathsf{X}, \mathsf{X})$ of all linear operators in X is not only a vector space but an *algebra*. $\mathscr{B}(\mathsf{X})$ is not *commutative* for $\dim \mathsf{X} \geqq 2$ since $TS = ST$ is in general not true. When $TS = ST$, T and S are said to *commute* (with each other). We have $T0 = 0T = 0$ and $T1 = 1T = T$ for every $T \in \mathscr{B}(\mathsf{X})$, where 1 denotes the *identity operator* (defined by $1u = u$ for every $u \in \mathsf{X}$). Thus 1 is the *unit element* of $\mathscr{B}(\mathsf{X})$[1]. The operators of the form $\alpha 1$ are called *scalar operators*[2] and in symbol will not be distinguished from the scalars α. A scalar operator commutes with every operator of $\mathscr{B}(\mathsf{X})$.

We write $TT = T^2$, $TTT = T^3$ and so on, and set $T^0 = 1$ by definition. We have

$$T^m T^n = T^{m+n}\,, \quad (T^m)^n = T^{mn}\,, \quad m, n = 0, 1, 2, \ldots \tag{3.17}$$

For any polynomial $p(z) = \alpha_0 + \alpha_1 z + \cdots + \alpha_n z^n$ in the indeterminate z, we define the operator

$$p(T) = \alpha_0 + \alpha_1 T + \cdots + \alpha_n T^n\,. \tag{3.18}$$

---

[1] Note that $1 \neq 0$ if (and only if) $\dim \mathsf{X} \geqq 1$.

[2] This should not be confused with the notion of scalar operators in the theory of spectral operators due to Dunford (see DUNFORD and SCHWARTZ [[1]]).

The mapping $p(z) \to p(T)$ is a *homomorphism* of the algebra of polynomials to $\mathscr{B}(\mathsf{X})$; this means that $p(z) + q(z) = r(z)$ or $p(z)\,q(z) = r(z)$ inplies $p(T) + q(T) = r(T)$ or $p(T)\,q(T) = r(T)$ respectively. In particular, it follows that $p(T)$ and $q(T)$ commute.

Problem 3.6. The operators $P_{jk} \in \mathscr{B}(\mathsf{X})$ given by (3.13) with $\mathsf{Y} = \mathsf{X}$, $y_j = x_j$ satisfy the relations

$$P_{jk}\,P_{ih} = \delta_{ki}\,P_{jh}\,, \quad j, k, i, h = 1, \ldots, N\,. \tag{3.19}$$

Problem 3.7. Set $\mathsf{R}_n = \mathsf{R}(T^n)$ and $\mathsf{N}_n = \mathsf{N}(T^n)$, $n = 0, 1, 2, \ldots$. The sequence $\{\mathsf{R}_n\}$ is nonincreasing and $\{\mathsf{N}_n\}$ is nondecreasing. There is a nonnegative integer $m \leqq \dim \mathsf{X}$ such that $\mathsf{R}_n \neq \mathsf{R}_{n+1}$ for $n < m$ and $\mathsf{R}_n = \mathsf{R}_{n+1}$ for $n \geqq m$.

If $T \in \mathscr{B}(\mathsf{X})$ is nonsingular, the inverse T^{-1} exists and belongs to $\mathscr{B}(\mathsf{X})$; we have

$$T^{-1}T = T\,T^{-1} = 1\,. \tag{3.20}$$

If T has a *left inverse* T' (that is, a $T' \in \mathscr{B}(\mathsf{X})$ such that $T'T = 1$), T has nullity zero, for $Tu = 0$ implies $u = T'Tu = 0$. If T has a *right inverse* T'' (that is, $T\,T'' = 1$), T has deficiency zero because every $u \in \mathsf{X}$ lies in $\mathsf{R}(T)$ by $u = T\,T''\,u$. If $\dim \mathsf{X}$ is finite, either of these facts implies that T is nonsingular and that $T' = T^{-1}$ or $T'' = T^{-1}$, respectively.

If S and T are nonsingular, so is TS and

$$(T\,S)^{-1} = S^{-1}\,T^{-1}\,. \tag{3.21}$$

For a nonsingular T, the negative powers T^{-n}, $n = 1, 2, \ldots$, can be defined by $T^{-n} = (T^{-1})^n$. In this case (3.17) is true for any integers m, n.

The following relations on determinants and traces follow directly from Problem 3.5:

$$\begin{aligned} &\det T\,S = (\det T)\,(\det S)\,,\\ &\operatorname{tr}(\alpha S + \beta T) = \alpha \operatorname{tr} S + \beta \operatorname{tr} T\,,\\ &\operatorname{tr} S\,T = \operatorname{tr} T\,S\,. \end{aligned} \tag{3.22}$$

Problem 3.8. The last formula of (3.22) is true even when $S \in \mathscr{B}(\mathsf{X}, \mathsf{Y})$ and $T \in \mathscr{B}(\mathsf{Y}, \mathsf{X})$ so that $S\,T \in \mathscr{B}(\mathsf{Y})$ and $T\,S \in \mathscr{B}(\mathsf{X})$.

Problem 3.8a. $\det(\alpha - TS) = \det(\alpha - ST)$ for $T, S \in \mathscr{B}(\mathsf{X})$. [hint: apply (3.22) to $(\alpha - TS)\,T = T\,(\alpha - ST)$.]

4. Projections. Nilpotents

Let M, N be two *complementary* linear manifolds of X; by this we mean that

$$\mathsf{X} = \mathsf{M} \oplus \mathsf{N}\,; \tag{3.23}$$

see § 1.3. Thus each $u \in \mathsf{X}$ can be uniquely expressed in the form $u = u' + u''$ with $u' \in \mathsf{M}$ and $u'' \in \mathsf{N}$. u' is called the *projection of u on* M

along N. If $v = v' + v''$ in the same sense, $\alpha u + \beta v$ has the projection $\alpha u' + \beta v'$ on M along N. If we set $u' = Pu$, it follows that P is a linear operator on X to itself. P is called the *projection operator* (or simply the *projection*) *on* M *along* N. $1 - P$ is the projection on N along M. We have $Pu = u$ if and only if $u \in \mathsf{M}$, and $Pu = 0$ if and only if $u \in \mathsf{N}$. The range of P is M and the null space of P is N. For convenience we often write $\dim P$ for $\dim \mathsf{M} = \dim \mathsf{R}(P)$. Since $Pu \in \mathsf{M}$ for every $u \in \mathsf{X}$, we have $PPu = Pu$, that is, P is *idempotent*:

$$P^2 = P . \tag{3.24}$$

Conversely, any idempotent operator P is a projection. In fact, set $\mathsf{M} = \mathsf{R}(P)$ and $\mathsf{N} = \mathsf{R}(1 - P)$. $u' \in \mathsf{M}$ implies that $u' = Pu$ for some u and therefore $Pu' = P^2 u = Pu = u'$. Similarly $u'' \in \mathsf{N}$ implies $Pu'' = 0$. Hence $u \in \mathsf{M} \cap \mathsf{N}$ implies that $u = Pu = 0$, so that $\mathsf{M} \cap \mathsf{N} = 0$. Each $u \in \mathsf{X}$ has the expression $u = u' + u''$ with $u' = Pu \in \mathsf{M}$ and $u'' = (1 - P) u \in \mathsf{N}$. This shows that P is the projection on M along N.

Problem 3.9. If P is a projection, we have

$$\operatorname{tr} P = \dim P . \tag{3.25}$$

The above results can be extended to the case in which there are several linear manifolds $\mathsf{M}_1, \ldots, \mathsf{M}_s$ such that

$$\mathsf{X} = \mathsf{M}_1 \oplus \cdots \oplus \mathsf{M}_s . \tag{3.26}$$

Each $u \in \mathsf{X}$ is then expressed in the form $u = u_1 + \cdots + u_s$, $u_j \in \mathsf{M}_j$, $j = 1, \ldots, s$, in a unique way. The operator P_j defined by $P_j u = u_j$ is the projection on M_j along $\mathsf{N}_j = \mathsf{M}_1 \oplus \cdots \oplus \mathsf{M}_{j-1} \oplus \mathsf{M}_{j+1} \oplus \cdots \oplus \mathsf{M}_s$. Furthermore, we have

$$\sum P_j = 1 , \tag{3.27}$$

$$P_k P_j = \delta_{jk} P_j . \tag{3.28}$$

Conversely, let $P_1, \ldots, P_s$ be operators satisfying the conditions (3.27) and (3.28)[1]. If we write $\mathsf{M}_j = \mathsf{R}(P_j)$, it is easily seen that (3.26) is satisfied and the P_j are the projections defined as above. In particular consider the case $s = 3$ and set $P = P_1 + P_2$. Then $P_1 = P_1 P = P P_1 = P P_1 P$; P_1 is a projection commuting with P and with $\mathsf{R}(P_1) \subset \mathsf{R}(P)$. Such a P_1 will be called a *subprojection of* P (a proper subprojection if $P_1 \neq P$ in addition), in symbol $P_1 \leqq P$.

A basis $\{x_j\}$ of X is said to be *adapted* to the decomposition (3.26) if the first several elements of $\{x_j\}$ belong to M_1, the following several ones belong to M_2, and so on. With respect to such a basis $\{x_j\}$, each P_j is

[1] Such a family is sometimes called a complete orthogonal family of projections. We do not use this term, to avoid a possible confusion with the notion of an orthogonal projection to be introduced in a unitary or Hilbert space.

represented by a diagonal matrix with diagonal elements equal to 0 or 1, the number of 1's being equal to $\dim \mathsf{M}_j$. Conversely, such a matrix always represents a projection.

For each linear manifold M of X, there is a complementary manifold N [such that (3.23) is true]. Thus every linear manifold has a projection on it. Such a projection is not unique, however.

A linear operator $T \in \mathscr{B}(\mathsf{X})$ is called a *nilpotent (operator)* if $T^n = 0$ for some positive integer n. A nilpotent is necessarily singular.

Let us consider the structure of a nilpotent T in more detail. Let n be such that $T^n = 0$ but $T^{n-1} \neq 0$ (we assume $\dim \mathsf{X} = N > 0$). Then $\mathsf{R}(T^{n-1}) \neq 0$; let $\{x_1^1, \ldots, x_{p_1}^1\}$ be a basis of $\mathsf{R}(T^{n-1})$. Each x_i^1 has the form $x_i^1 = T^{n-1} x_i^n$ for some $x_i^n \in \mathsf{X}$, $i = 1, \ldots, p_1$. If $n > 1$, set $T^{n-2} x_i^n = x_i^2$ so that $T x_i^2 = x_i^1$. The vectors x_i^k, $k = 1, 2$, $i = 1, \ldots, p_1$, belong to $\mathsf{R}(T^{n-2})$ and are linearly independent; in fact $\sum \alpha_i x_i^2 + \sum \beta_i x_i^1 = 0$ implies $\sum \alpha_i x_i^1 = 0$ on application of T and hence $\alpha_i = 0$ for all i, hence $\sum \beta_i x_i^1 = 0$ and $\beta_i = 0$ for all i. Let us enlarge the family $\{x_i^k\}$ to a basis of $\mathsf{R}(T^{n-2})$ by adding, if necessary, new vectors $x_{p_1+1}^2, \ldots, x_{p_2}^2$; here we can arrange that $T x_i^2 = 0$ for $i > p_1$.

If $n > 2$ we can proceed in the same way. Finally we arrive at a basis $\{x_j^k\}$ of X with the following properties: $k = 1, \ldots, n$, $j = 1, \ldots, p_k$, $p_1 \leqq p_2 \leqq \cdots \leqq p_n$,

$$T x_j^k = \begin{cases} x_j^{k-1}, & 1 \leqq j \leqq p_{k-1}, \\ 0, & p_{k-1} + 1 \leqq j \leqq p_k, \end{cases} \tag{3.29}$$

where we set $p_0 = 0$.

If we arrange the basis $\{x_j^k\}$ in the order $\{x_1^1, \ldots, x_1^n, x_2^1, \ldots, x_2^n, \ldots\}$, the matrix of T with respect to this basis takes the form

$$\left(\begin{array}{ccccc|cccc|c} 0 & 1 & & & & & & & & \\ & 0 & 1 & & & & & & & \\ & & \ddots & \ddots & & & & & & \\ & & & 0 & 1 & & & & & \\ & & & & 0 & & & & & \\ \hline & & & & & 0 & 1 & & & \\ & & & & & & 0 & 1 & & \\ & & & & & & & \cdot & 1 & \\ & & & & & & & & 0 & \\ \hline & & & & & & & & & \ddots \end{array}\right). \quad \text{(all unspecified elements are zero)} \tag{3.30}$$

Problem 3.10. If T is nilpotent, then $T^N = 0$ for $N = \dim \mathsf{X}$.

Problem 3.11. If T is nilpotent, then $\operatorname{tr} T = 0$ and $\det(1 + T) = 1$.

5. Invariance. Decomposition

A linear manifold M is said to be *invariant* under an operator $T \in \mathscr{B}(\mathsf{X})$ if $T\mathsf{M} \subset \mathsf{M}$. In this case T *induces* a linear operator T_M on M to M, defined by $T_\mathsf{M} u = Tu$ for $u \in \mathsf{M}$. T_M is called the *part of T in* M.

Problem 3.12. $\mathsf{R}_n = \mathsf{R}(T^n)$, $n = 0, 1, 2, \ldots$, are invariant under T. If m is defined as in Problem 3.7, the part of T in R_n is singular if $n < m$ and nonsingular if $n \geqq m$.

Problem 3.13. If M is invariant under T, M is also invariant under $p(T)$ for any polynomial $p(z)$, and $p(T)_{\mathsf{M}} = p(T_{\mathsf{M}})$.

If there are two invariant linear manifolds M, N for T such that $\mathsf{X} = \mathsf{M} \oplus \mathsf{N}$, T is said to be *decomposed* (or *reduced*) by the pair M, N. More generally, T is said to be decomposed by the set of linear manifolds $\mathsf{M}_1, \ldots, \mathsf{M}_s$ if (3.26) is satisfied and all the M_j are invariant under T [or we say that T is decomposed according to the decomposition (3.26)]. In this case T is completely described by its parts T_{M_j}, $j = 1, \ldots, s$. T is called the *direct sum* of the T_{M_j}. If $\{P_j\}$ is the set of projections corresponding to (3.26), T commutes with each P_j. In fact we have, successively, $P_j u \in \mathsf{M}_j$, $T P_j u \in \mathsf{M}_j$, $P_k T P_j u = \delta_{jk} T P_j u$, and the addition of the last equalities for $j = 1, \ldots, s$ gives $P_k T u = T P_k u$ or $P_k T = T P_k$. Conversely, it is easy to see that T is decomposed by $\mathsf{M}_1, \ldots, \mathsf{M}_s$ if T commutes with all the P_j.

If we choose a basis $\{x_j\}$ adapted to the decomposition (3.26), T is represented by a matrix which has non-zero elements only in s smaller submatrices along the diagonal (which are the matrices of the T_{M_j}). Thus the matrix of T is the direct sum of the matrices of the T_{M_j}.

Problem 3.14. With the above notations, we have

$$\det T = \prod_j \det T_{\mathsf{M}_j}, \quad \operatorname{tr} T = \sum_j \operatorname{tr} T_{\mathsf{M}_j}. \tag{3.31}$$

Remark 3.15. The operator $P_j T = T P_j = P_j T P_j$ coincides with T and also with T_{M_j} when applied to a $u \in \mathsf{M}_j$; it is sometimes identified with T_{M_j} when there is no possibility of misunderstanding.

6. The adjoint operator

Let $T \in \mathscr{B}(\mathsf{X}, \mathsf{Y})$. For each $g \in \mathsf{Y}^*$ and $u \in \mathsf{X}$, the scalar product $(g, T u)$ is defined and is a semilinear form in u. Therefore, it can be written as $f[u] = (f, u)$ with an $f \in \mathsf{X}^*$. Since f is determined by g, a function T^* on Y^* to X^* is defined by setting $f = T^* g$. Thus the defining equation of T^* is

$$(T^* g, u) = (g, T u), \quad g \in \mathsf{Y}^*, \quad u \in \mathsf{X}. \tag{3.32}$$

T^* is a linear operator on Y^* to X^*, that is, $T^* \in \mathscr{B}(\mathsf{Y}^*, \mathsf{X}^*)$. In fact, we have $(T^*(\alpha_1 g_1 + \alpha_2 g_2), u) = (\alpha_1 g_1 + \alpha_2 g_2, T u) = \alpha_1 (g_1, T u) + \alpha_2 (g_2, T u) = \alpha_1 (T^* g_1, u) + \alpha_2 (T^* g_2, u) = (\alpha_1 T^* g_1 + \alpha_2 T^* g_2, u)$ so that $T^*(\alpha_1 g_1 + \alpha_2 g_2) = \alpha_1 T^* g_1 + \alpha_2 T^* g_2$. T^* is called the *adjoint (operator)* of T.

The operation * has the following properties:

$$(3.33)\qquad (\alpha S + \beta T)^* = \bar{\alpha} S^* + \bar{\beta} T^*, \quad (TS)^* = S^* T^* .$$

In the second formula it is assumed that $T \in \mathscr{B}(\mathsf{Y}, \mathsf{Z})$ and $S \in \mathscr{B}(\mathsf{X}, \mathsf{Y})$ so that TS is defined and belongs to $\mathscr{B}(\mathsf{X}, \mathsf{Z})$; note that $S^* \in \mathscr{B}(\mathsf{Y}^*, \mathsf{X}^*)$ and $T^* \in \mathscr{B}(\mathsf{Z}^*, \mathsf{Y}^*)$ so that $S^* T^* \in \mathscr{B}(\mathsf{Z}^*, \mathsf{X}^*)$. The proof of (3.33) is simple; for example, the second formula follows from $((TS)^* h, u) = (h, TSu) = (T^* h, Su) = (S^* T^* h, u)$ which is valid for all $h \in \mathsf{Z}^*$ and $u \in \mathsf{X}$.

Problem 3.16. $0^* = 0$, $1^* = 1$ (the 0 on the left is the zero of $\mathscr{B}(\mathsf{X}, \mathsf{Y})$ while the 0 on the right is the zero of $\mathscr{B}(\mathsf{Y}^*, \mathsf{X}^*)$; similarly for the second equality, in which we must set $\mathsf{Y} = \mathsf{X}$).

If $T \in \mathscr{B}(\mathsf{X}, \mathsf{Y})$, we have $T^* \in \mathscr{B}(\mathsf{Y}^*, \mathsf{X}^*)$ and $T^{**} \in \mathscr{B}(\mathsf{X}^{**}, \mathsf{Y}^{**})$. If we identify X^{**} and Y^{**} with X and Y respectively (see § 2.6), it follows from (3.32) that

$$(3.34)\qquad T^{**} = T .$$

If we take bases $\{x_k\}$ and $\{y_j\}$ in X and Y respectively, an operator $T \in \mathscr{B}(\mathsf{X}, \mathsf{Y})$ is represented by a matrix (τ_{jk}) according to (3.5) or (3.6). If $\{e_k\}$ and $\{f_j\}$ are the adjoint bases of X^* and Y^*, respectively, the operator $T^* \in \mathscr{B}(\mathsf{Y}^*, \mathsf{X}^*)$ can similarly be represented by a matrix (τ^*_{kj}). These matrix elements are given by $\tau_{jk} = (T x_k, f_j)$ and $\tau^*_{kj} = (T^* f_j, x_k) = (f_j, T x_k)$ in virtue of (3.10). Thus

$$(3.35)\qquad \tau^*_{kj} = \overline{\tau_{jk}}, \quad \begin{array}{l} k = 1, \ldots, N = \dim \mathsf{X}, \\ j = 1, \ldots, M = \dim \mathsf{Y}, \end{array}$$

and T and T^* are represented by mutually *adjoint* (Hermitian conjugate) matrices with respect to the bases which are adjoint to each other.

Problem 3.17. If $T \in \mathscr{B}(\mathsf{X})$, we have

$$(3.36)\qquad \det T^* = \overline{\det T}, \quad \operatorname{tr} T^* = \overline{\operatorname{tr} T} .$$

Let $T \in \mathscr{B}(\mathsf{X}, \mathsf{Y})$. A $g \in \mathsf{Y}^*$ belongs to the annihilator of $\mathsf{R}(T)$ if and only if $(g, Tu) = 0$ for all $u \in \mathsf{X}$. (3.32) shows that this is equivalent to $T^* g = 0$. Thus *the annihilator of the range of T is identical with the null space of T^**. In view of (3.34), the same is true when T and T^* are exchanged. In symbol, we have

$$(3.37)\qquad \mathsf{N}(T^*) = \mathsf{R}(T)^\perp, \quad \mathsf{N}(T) = \mathsf{R}(T^*)^\perp .$$

It follows that [see (3.2), (3.3) and (2.17)]

$$(3.38)\qquad \operatorname{nul} T^* = \operatorname{def} T, \quad \operatorname{nul} T = \operatorname{def} T^*, \quad \operatorname{rank} T^* = \operatorname{rank} T .$$

If in particular $\mathsf{Y} = \mathsf{X}$, (3.38) shows that T^* is nonsingular if and only if T is; in this case we have

$$(3.39)\qquad (T^*)^{-1} = (T^{-1})^* .$$

For the proof it suffices to note that $T^*(T^{-1})^* = (T^{-1}T)^* = 1^* = 1$.

Problem 3.18. If $T \in \mathscr{B}(\mathsf{X})$, we have

$$\operatorname{nul} T^* = \operatorname{nul} T, \quad \operatorname{def} T^* = \operatorname{def} T. \tag{3.40}$$

If $P \in \mathscr{B}(\mathsf{X})$ is a projection, the adjoint $P^* \in \mathscr{B}(\mathsf{X}^*)$ is likewise a projection, for $P^2 = P$ implies $P^{*2} = P^*$. The decompositions of the spaces X and X^*

$$\mathsf{X} = \mathsf{M} \oplus \mathsf{N}, \quad \mathsf{M} = \mathsf{R}(P), \quad \mathsf{N} = \mathsf{R}(1 - P), \tag{3.41}$$

$$\mathsf{X}^* = \mathsf{M}^* \oplus \mathsf{N}^*, \; \mathsf{M}^* = \mathsf{R}(P^*), \; \mathsf{N}^* = \mathsf{R}(1 - P^*), \tag{3.42}$$

are related to each other through the following equalities:

$$\mathsf{N}^* = \mathsf{M}^\perp, \quad \mathsf{M}^* = \mathsf{N}^\perp, \quad \dim \mathsf{M}^* = \dim \mathsf{M}, \quad \dim \mathsf{N}^* = \dim \mathsf{N}, \tag{3.43}$$

as is seen from (3.37) and (3.40).

Similar results hold when there are several projections. If $\{P_j\}$ is a set of projections in X satisfying (3.27–3.28), $\{P_j^*\}$ is a similar set of projections in X^*. The ranges $\mathsf{M}_j = \mathsf{R}(P_j)$, $\mathsf{M}_j^* = \mathsf{R}(P_j^*)$ are related by

$$\dim \mathsf{M}_j^* = \dim \mathsf{M}_j, \quad j = 1, 2, \ldots, \tag{3.44}$$

$$\mathsf{M}_1^* = (\mathsf{M}_2 \oplus \cdots)^\perp, \quad \mathsf{M}_1^\perp = (\mathsf{M}_2^* \oplus \cdots)^\perp, \quad \text{etc.} \tag{3.45}$$

Problem 3.19. Let $\{x_j\}$ be a basis of X adapted to the decomposition $\mathsf{X} = \mathsf{M}_1 \oplus \oplus \cdots \oplus \mathsf{M}_s$, and let $\{e_j\}$ be the adjoint basis of X^*. Then $\{e_j\}$ is adapted to the decomposition $\mathsf{X}^* = \mathsf{M}_1^* \oplus \cdots \oplus \mathsf{M}_s^*$. For any $u \in \mathsf{X}$ we have

$$P_j u = \sum_{r=1}^{m_j} (u, e_{jr}) \, x_{jr}, \tag{3.46}$$

where $\{x_{j1}, \ldots, x_{jm_j}\}$ is the part of $\{x_j\}$ belonging to M_j and $m_j = \dim \mathsf{M}_j$.

§ 4. Analysis with operators

1. Convergence and norms for operators

Since the set $\mathscr{B}(\mathsf{X}, \mathsf{Y})$ of all linear operators on X to Y is an MN-dimensional vector space, where $N = \dim \mathsf{X} < \infty$ and $M = \dim \mathsf{Y} < \infty$, the notion of convergence of a sequence $\{T_n\}$ of operators of $\mathscr{B}(\mathsf{X}, \mathsf{Y})$ is meaningful as in the case of a sequence of vectors of X. If we introduce the matrix representation (τ_{njk}) of T_n with respect to fixed bases $\{x_k\}$, $\{y_j\}$ of X, Y, respectively, $T_n \to T$ is equivalent to $\tau_{njk} \to \tau_{jk}$ for each j, k, for the τ_{njk} are the coefficients of T_n for the basis $\{P_{jk}\}$ of $\mathscr{B}(\mathsf{X}, \mathsf{Y})$ (see § 3.2). But the τ_{njk} are the coefficients of $T x_k$ with respect to the basis $\{y_j\}$; hence $T_n \to T$ is equivalent to $T_n x_k \to T x_k$ for all k and therefore to $T_n u \to T u$ for all $u \in \mathsf{X}$. This could have been used as the definition of $T_n \to T$.

As it was convenient to express the convergence of vectors by means of a norm, so it is with operators. But an operator-norm is usually introduced only in correlation with the vector-norms. More precisely, when X and Y are normed spaces $\mathscr{B}(\mathsf{X}, \mathsf{Y})$ is defined to be a normed space with the norm given by

$$(4.1) \qquad \|T\| = \sup_{0 \neq u \in \mathsf{X}} \frac{\|Tu\|}{\|u\|} = \sup_{\|u\|=1} \|Tu\| = \sup_{\|u\| \leqq 1} \|Tu\|, \quad T \in \mathscr{B}(\mathsf{X}, \mathsf{Y}).$$

The equality of the various expressions in (4.1) is easily verified[1]. We can replace "sup" by "max" in (4.1) because the set of u with $\|u\| = 1$ or $\|u\| \leqq 1$ is compact (see a similar remark for the norm of an $f \in \mathsf{X}^*$ in § 2.4); this shows that $\|T\|$ is finite. It is easy to verify that $\|T\|$ defined on $\mathscr{B}(\mathsf{X}, \mathsf{Y})$ by (4.1) satisfies the conditions (1.18) of a norm. Hence it follows that $T_n \to T$ is equivalent to $\|T_n - T\| \to 0$. A necessary and sufficient condition that $\{T_n\}$ converge to some T is given by the Cauchy condition $\|T_n - T_m\| \to 0$, $n, m \to \infty$.

Another convenient expression for $\|T\|$ is[2]

$$(4.2) \qquad \|T\| = \sup_{\substack{0 \neq u \in \mathsf{X} \\ 0 \neq f \in \mathsf{Y}^*}} \frac{|(Tu, f)|}{\|f\| \, \|u\|} = \sup_{\substack{\|u\|=1 \\ \|f\|=1}} |(Tu, f)|.$$

The equivalence of (4.2) with (4.1) follows from (2.26).

If we introduce different norms in the given vector spaces X and Y, $\mathscr{B}(\mathsf{X}, \mathsf{Y})$ acquires different norms accordingly. As in the case of norms in X, however, all these norms in $\mathscr{B}(\mathsf{X}, \mathsf{Y})$ are equivalent, in the sense that for any two norms $\| \ \|$ and $\| \ \|'$, there are positive constants α', β' such that

$$(4.3) \qquad \alpha' \|T\| \leqq \|T\|' \leqq \beta' \|T\|.$$

This is a special case of (1.20) applied to $\mathscr{B}(\mathsf{X}, \mathsf{Y})$ regarded as a normed space. Similarly, the inequalities (1.21) and (1.22) give the following inequalities:

$$(4.4) \qquad |\tau_{jk}| \leqq \gamma \|T\|, \quad j = 1, \ldots, M; \; k = 1, \ldots, N,$$

$$(4.5) \qquad \|T\| \leqq \gamma' \max |\tau_{jk}|,$$

where (τ_{jk}) is the matrix of T with respect to the bases of X and Y. The constants γ and γ' depend on these bases and the norm employed, but are independent of the operator T.

As in the case of vectors, $\alpha S + \beta T$ is a continuous function of the scalars α, β and the operators $S, T \in \mathscr{B}(\mathsf{X}, \mathsf{Y})$, and $\|T\|$ is a continuous function of T. As a new feature of the norm of operators, we should note the inequality

[1] The second and third members of (4.1) do not make sense if $\dim \mathsf{X} = 0$; in this case we have simply $\|T\| = 0$.

[2] Here we assume $\dim \mathsf{X} \geqq 1$, $\dim \mathsf{Y} \geqq 1$.

$$(4.6) \qquad \|TS\| \leqq \|T\|\,\|S\| \quad \text{for} \quad T \in \mathscr{B}(\mathsf{Y}, \mathsf{Z}) \quad \text{and} \quad S \in \mathscr{B}(\mathsf{X}, \mathsf{Y}) .$$

This follows from $\|TSu\| \leqq \|T\|\,\|Su\| \leqq \|T\|\,\|S\|\,\|u\|$; note also that (4.6) would not be valid if we chose arbitrary norms in $\mathscr{B}(\mathsf{X}, \mathsf{Y})$ etc. regarded simply as vector spaces.

Problem 4.1. $\|1\| = 1$ ($1 \in \mathscr{B}(\mathsf{X})$ is the identity operator, $\dim \mathsf{X} > 0$). If $P \in \mathscr{B}(\mathsf{X})$ is a projection and $P \neq 0$, then $\|P\| \geqq 1$.

TS is a continuous function of S and T. In other words, $T_n \to T$ and $S_n \to S$ imply $T_n S_n \to TS$. The proof is similar to (2.28); it suffices to use (4.6). In the same way, it can be shown that Tu is a continuous function of T and u. In particular $u_n \to u$ implies $Tu_n \to Tu$. In this sense a linear operator T is a continuous function. It is permitted, for example, to operate with T term by term on a convergent series of vectors:

$$(4.7) \qquad T\Big(\sum_n u_n\Big) = \sum_n T u_n .$$

If X is a normed space, $\mathscr{B}(\mathsf{X}) = \mathscr{B}(\mathsf{X}, \mathsf{X})$ is a *normed algebra* (or *ring*) with the norm given by (4.2). In particular, (4.6) is true for $T, S \in \mathscr{B}(\mathsf{X})$.

If $T \in \mathscr{B}(\mathsf{X}, \mathsf{Y})$, then $T^* = \mathscr{B}(\mathsf{Y}^*, \mathsf{X}^*)$ and

$$(4.8) \qquad \|T^*\| = \|T\| .$$

This follows from (4.2), according to which $\|T^*\| = \sup |(T^* f, u)| = \sup |(f, Tu)| = \|T\|$ where $u \in \mathsf{X}^{**} = \mathsf{X}$, $\|u\| = 1$ and $f \in \mathsf{X}^*$, $\|f\| = 1$.

Problem 4.1a. rank T is a lower semicontinuous function of T. nul T and def T are upper semicontinuous. In other words, $\lim T_n = T$ implies $\liminf \operatorname{rank} T_n \geqq \operatorname{rank} T$, $\limsup \operatorname{nul} T_n \leqq \operatorname{nul} T$, $\limsup \operatorname{def} T_n \leqq \operatorname{def} T$. [hint: if rank $T = r$, there are r vectors u_j such that $T u_j$ are linearly independent. Then the $T_n u_j$ are linearly independent for sufficiently large n (see Problem 2.8a).]

2. The norm of T^n

As an example of the use of the norm and also with a view to later applications, we consider the norm $\|T^n\|$ for $T \in \mathscr{B}(\mathsf{X})$. It follows from (4.6) that

$$(4.9) \qquad \|T^{m+n}\| \leqq \|T^m\|\,\|T^n\| , \quad \|T^n\| \leqq \|T\|^n , \quad m, n = 0, 1, 2, \ldots .$$

We shall show that $\lim_{n\to\infty} \|T^n\|^{1/n}$ *exists and is equal to* $\inf_{n=1,2,\ldots} \|T^n\|^{1/n}$. This limit is called the *spectral radius* of T and will be denoted by $\operatorname{spr} T$. As is seen later, $\operatorname{spr} T$ is independent of the norm employed in its definition.

Set $a_n = \log\|T^n\|$. What is to be proved is that

$$(4.10) \qquad a_n/n \to b \equiv \inf_{n=1,2,\ldots} a_n/n .$$

The inequality (4.9) gives

$$(4.11) \qquad a_{m+n} \leqq a_m + a_n .$$

(Such a sequence $\{a_n\}$ is said to be *subadditive.*) For a fixed positive integer m, set $n = mq + r$, where q, r are nonnegative integers with $0 \leqq r < m$. Then (4.11) gives $a_n \leqq q\, a_m + a_r$ and

$$\frac{a_n}{n} \leqq \frac{q}{n} a_m + \frac{1}{n} a_r .$$

If $n \to \infty$ for a fixed m, $q/n \to 1/m$ whereas r is restricted to one of the numbers $0, 1, \ldots, m-1$. Hence $\limsup_{n\to\infty} a_n/n \leqq a_m/m$. Since m is arbitrary, we have $\limsup a_n/n \leqq b$. On the other hand, we have $a_n/n \geqq b$ and so $\liminf a_n/n \geqq b$. This proves (4.10)[1].

Remark 4.2. The above result may lead one to conjecture that $\|T^n\|^{1/n}$ is monotone nonincreasing. This is not true, however, as is seen from the following example. Let $\mathsf{X} = C^2$ with the norm given by (1.17) (X is a two-dimensional unitary space, see § 6). Let T be given by the matrix

$$T = \begin{pmatrix} 0 & a^2 \\ b^2 & 0 \end{pmatrix}, \quad a > b > 0 .$$

It is easily seen that (1 is the unit matrix)

$$T^{2n} = a^{2n} b^{2n}\, 1\,, \qquad T^{2n+1} = a^{2n} b^{2n}\, T\,, \quad \|T^{2n}\|^{1/2n} = a\, b\,,$$
$$\|T\| = a^2\,, \quad \|T^{2n+1}\|^{1/(2n+1)} = a\, b\, (a/b)^{1/(2n+1)} > a\, b\,.$$

Next let us consider $\|T^{-1}\|$ and deduce an inequality estimating $\|T^{-1}\|$ in terms of $\|T\|$ and $\det T$, assuming that T is nonsingular. The relation $Tu = v$ is expressed by the linear equations (3.6). Solving these equations for ξ_k, we obtain a fractional expression of which the denominator is $\det T$ and the numerator is a linear combination of the η_k with coefficients that are equal to minors of the matrix (τ_{jk}). These minors are polynomials in the τ_{jk} of degree $N-1$ where $N = \dim \mathsf{X}$. In virtue of the inequalities $\|u\| \leqq \gamma' \max |\xi_j|$ [see (1.22)], $|\tau_{jk}| \leqq \gamma'' \|T\|$ [see (4.4)] and $|\eta_j| \leqq \gamma''' \|v\|$ [see (1.21)], it follows that there is a constant γ such that $\|u\| \leqq \gamma \|v\|\, \|T\|^{N-1}/|\det T|$ or

$$\|T^{-1}\| \leqq \gamma \frac{\|T\|^{N-1}}{|\det T|} . \tag{4.12}$$

The constant γ is independent of T, depending only on the norm employed[2].

3. Examples of norms

Since the norm $\|T\|$ of an operator T is determined in correlation with the norms adopted in the domain and the range spaces X, Y, there is not so much freedom in the choice of $\|T\|$ as in the choice of the norms for vectors. For the same reason it is not always easy to compute the exact value of $\|T\|$. It is often required, however, to estimate an upper bound of $\|T\|$. We shall illustrate by examples how such an estimate is obtained.

---

[1] See Pólya and Szegö [[1]], p. 17. Cf. also Hille and Phillips [[1]], pp. 124, 244.
[2] We can set $\gamma = 1$ if X is a unitary space; see T. Kato [13].

Most commonly used among the norms of vectors is the *p-norm* defined by

$$\|u\| = \|u\|_p = \Big(\sum_j |\xi_j|^p\Big)^{1/p} \tag{4.13}$$

with a fixed $p \geqq 1$, where the ξ_j are the coefficients of u with respect to a fixed basis $\{x_j\}$ (which will be called the *canonical basis*). The conditions (1.18) for a norm are satisfied by this p-norm [the third condition of (1.18) is known as the Minkowski inequality[1]]. The special cases $p = 1$ and 2 were mentioned before; see (1.16) and (1.17). The norm $\|u\| = \max|\xi_j|$ given by (1.15) can be regarded as the limiting case of (4.13) for $p = \infty$.

Suppose now that the p-norm is given in X and Y with the same p. We shall estimate the corresponding norm $\|T\|$ of an operator T on X to Y in terms of the matrix (τ_{jk}) of T with respect to the canonical bases of X and Y. If $v = Tu$, the coefficients ξ_k and η_j of u and v, respectively, are related by the equations (3.6). Let

$$\tau'_j = \sum_k |\tau_{jk}|\,, \quad \tau''_k = \sum_j |\tau_{jk}|\,, \tag{4.14}$$

be the row sums and the column sums of the matrix $(|\tau_{jk}|)$. (3.6) then gives

$$\frac{|\eta_j|}{\tau'_j} \leqq \sum_k \frac{|\tau_{jk}|}{\tau'_j}\,|\xi_k|\,.$$

Since the nonnegative numbers $|\tau_{jk}|/\tau'_j$ for $k = 1, \ldots, N$ with a fixed j have the sum 1, the right member is a weighted average of the $|\xi_k|$. Since λ^p is a convex function of $\lambda \geqq 0$, the p-th power of the right member does not exceed the weighted average of the $|\xi_k|^p$ with the same weights[2]. Thus

$$\left(\frac{|\eta_j|}{\tau'_j}\right)^p \leqq \sum_k \frac{|\tau_{jk}|}{\tau'_j}\,|\xi_k|^p\,,$$

and we have successively

$$|\eta_j|^p \leqq \tau'^{\,p-1}_j \sum_k |\tau_{jk}|\,|\xi_k|^p \leqq \big(\max_j \tau'_j\big)^{p-1} \sum_k |\tau_{jk}|\,|\xi_k|^p\,,$$

$$\|v\|^p = \sum_j |\eta_j|^p \leqq \big(\max_j \tau'_j\big)^{p-1} \sum_k \tau''_k\,|\xi_k|^p \leqq \big(\max_j \tau'_j\big)^{p-1}\big(\max_k \tau''_k\big)\|u\|^p\,,$$

hence

$$\|Tu\| = \|v\| \leqq \big(\max_j \tau'_j\big)^{1-\frac{1}{p}}\big(\max_k \tau''_k\big)^{\frac{1}{p}}\|u\|\,. \tag{4.15}$$

This shows that[3]

$$\|T\| \leqq \big(\max_j \tau'_j\big)^{1-\frac{1}{p}}\big(\max_k \tau''_k\big)^{\frac{1}{p}}. \tag{4.16}$$

If $p = 1$, the first factor on the right of (4.16) is equal to 1 and does not depend on the τ'_j. On letting $p \to \infty$, it is seen that (4.16) is true also for $p = \infty$; then the second factor on the right is 1 and does not depend on the τ''_k.

Problem 4.3. If (τ_{jk}) is a diagonal matrix ($\tau_{jk} = 0$ for $j \neq k$), we have for any p

$$\|T\| \leqq \max_j |\tau_{jj}|\,. \tag{4.17}$$

---

[1] The proof may be found in any textbook on real analysis. See e. g. Hardy, Littlewood and Pólya [1], p. 31; Royden [1], p. 97.

[2] For convex functions, see e. g. Hardy, Littlewood and Pólya [1], p. 70.

[3] Actually this is a simple consequence of the convexity theorem of M. Riesz (see Hardy, Littlewood and Pólya [1], p. 203).

4. Infinite series of operators

The convergence of an infinite series of operators $\sum T_n$ can be defined as for infinite series of vectors and need not be repeated here. Similarly, the *absolute convergence* of such a series means that the series $\sum \|T_n\|$ is convergent for some (and hence for any) norm $\| \ \|$. In this case $\sum T_n$ is convergent with $\|\sum T_n\| \leqq \sum \|T_n\|$.

Owing to the possibility of multiplication of operators, there are certain formulas for series of operators that do not exist for vectors. For example, we have

$$S(\sum T_n) = \sum S T_n , \quad (\sum T_n) S = \sum T_n S , \tag{4.18}$$

whenever $\sum T_n$ is convergent and the products are meaningful. This follows from the continuity of $S T$ as function of S and T. Two absolutely convergent series can be multiplied term by term, that is

$$(\sum S_m)(\sum T_n) = \sum S_m T_n \tag{4.19}$$

if the products are meaningful. Here the order of the terms on the right is arbitrary (or it may be regarded as a double series). The proof is not essentially different from the case of numerical series and may be omitted.

Example 4.4 (Exponential function)

$$e^{tT} = \exp(tT) = \sum_{n=0}^{\infty} \frac{1}{n!} t^n T^n , \quad T \in \mathscr{B}(\mathsf{X}) . \tag{4.20}$$

This series is absolutely convergent for every complex number t, for the n-th term is majorized by $|t|^n \|T\|^n/n!$ in norm. We have

$$\|e^{tT}\| \leqq e^{|t| \, \|T\|} . \tag{4.21}$$

Furthermore, we have

$$e^{t(T+S)} = e^{tT} e^{tS} = e^{tS} e^{tT} \quad \text{if} \quad T S = S T. \tag{4.21a}$$

This can be verified by straightforward computation based on (4.20). An alternative proof is given by using $(d/dt)\, e^{tT} e^{tS} = (T+S)\, e^{tT} e^{tS}$ [see (4.31) below]. Thus $(d/dt)^n e^{tT} e^{tS} = (T+S)^n e^{tT} e^{tS}$, $n = 1, 2, \ldots$, and (4.21 a) follows as the Taylor expansion for $e^{tT} e^{tS}$. As special cases of (4.21 a), we obtain

$$e^{(t+s)T} = e^{tT} e^{sT}, \quad e^{t(T+c)} = e^{ct} e^{tT}, \tag{4.21b}$$

where c is a scalar.

(4.21 b) shows that $\{e^{tT}\}$ forms a *group* of operators, where t varies over all complex (or real) numbers. T is called the (infinitesimal) *generator* of the group $\{e^{tT}\}$; it is uniquely determined from the group by

$$T = \lim_{t \to 0} t^{-1} (e^{tT} - 1). \tag{4.21c}$$

Example 4.5 (Neumann series)

$$(4.22) \qquad (1-T)^{-1} = \sum_{n=0}^{\infty} T^n, \quad \|(1-T)^{-1}\| \leqq (1-\|T\|)^{-1}, \quad T \in \mathscr{B}(\mathsf{X}).$$

This series is absolutely convergent for $\|T\| < 1$ in virtue of $\|T^n\| \leqq \|T\|^n$. Denoting the sum by S, we have $TS = ST = S - 1$ by term by term multiplication. Hence $(1-T)S = S(1-T) = 1$ and $S = (1-T)^{-1}$. It follows that *an operator* $R \in \mathscr{B}(\mathsf{X})$ *is nonsingular if* $\|1 - R\| < 1$.

It should be noted that whether or not $\|T\| < 1$ (or $\|1 - R\| < 1$) may depend on the norm employed in X; it may well happen that $\|T\| < 1$ holds for some norm but not for another.

Problem 4.6. The series (4.22) is absolutely convergent if $\|T^m\| < 1$ for some positive integer m or, equivalently, if $\operatorname{spr} T < 1$ (for $\operatorname{spr} T$ see § 4.2), and the sum is again equal to $(1-T)^{-1}$.

In the so-called *iteration method* in solving the linear equation $(1-T)u = v$ for u, the partial sums $S_n = \sum_{k=0}^{n} T^k$ are taken as approximations for $S = (1-T)^{-1}$ and $u_n = S_n v$ as approximations for the solution $u = Sv$. The errors incurred in such approximations can be estimated by

$$(4.23) \qquad \|S - S_n\| = \left\| \sum_{k=n+1}^{\infty} T^k \right\| \leqq \sum_{k=n+1}^{\infty} \|T\|^k = \frac{\|T\|^{n+1}}{1 - \|T\|}.$$

For $n = 0$ (4.23) gives $\|(1-T)^{-1} - 1\| \leqq \|T\| (1 - \|T\|)^{-1}$. With $R = 1 - T$, this shows that $R \to 1$ implies $R^{-1} \to 1$. In other words, R^{-1} is a continuous function of R at $R = 1$. This is a special case of the theorem that T^{-1} *is a continuous function of* T. More precisely, if $T \in \mathscr{B}(\mathsf{X}, \mathsf{Y})$ is nonsingular, any $S \in \mathscr{B}(\mathsf{X}, \mathsf{Y})$ with sufficiently small $\|S - T\|$ is also nonsingular, and $\|S^{-1} - T^{-1}\| \to 0$ for $\|T - S\| \to 0$. In particular, *the set of all nonsingular elements of* $\mathscr{B}(\mathsf{X}, \mathsf{Y})$ *is open*. [Of course X and Y must have the same dimension if there exist nonsingular elements of $\mathscr{B}(\mathsf{X}, \mathsf{Y})$.]

To see this we set $A = S - T$ and assume that $\|A\| < 1/\|T^{-1}\|$. Then $\|A T^{-1}\| \leqq \|A\| \, \|T^{-1}\| < 1$, and so $1 + A T^{-1}$ is a nonsingular operator of $\mathscr{B}(\mathsf{Y})$ by the above result. Since $S = T + A = (1 + A T^{-1}) T$, S is also nonsingular with $S^{-1} = T^{-1}(1 + A T^{-1})^{-1}$.

Using the estimates for $\|(1 + A T^{-1})^{-1}\|$ and $\|(1 + A T^{-1})^{-1} - 1\|$ given by (4.22) and (4.23), we obtain the following estimates for $\|S^{-1}\|$ and $\|S^{-1} - T^{-1}\|$:

$$(4.24) \qquad \|S^{-1}\| \leqq \frac{\|T^{-1}\|}{1 - \|A\| \, \|T^{-1}\|}, \quad \|S^{-1} - T^{-1}\| \leqq \frac{\|A\| \, \|T^{-1}\|^2}{1 - \|A\| \, \|T^{-1}\|}$$
$$\text{for } S = T + A, \quad \|A\| < 1/\|T^{-1}\|.$$

Remark 4.7. We assumed above that $\|A\| < \|T^{-1}\|$ to show the existence of S^{-1}. This condition can be weakened if $X = Y$ and $TS = ST$. In this case A commutes with T and hence with T^{-1}. Therefore

$$\begin{aligned} \operatorname{spr} A T^{-1} &= \lim \|(A T^{-1})^n\|^{1/n} = \lim \|A^n T^{-n}\|^{1/n} \leqq \\ &\leqq [\lim \|A^n\|^{1/n}] [\lim \|T^{-n}\|^{1/n}] = (\operatorname{spr} A) (\operatorname{spr} T^{-1}) . \end{aligned} \tag{4.25}$$

It follows that $S^{-1} = T^{-1}(1 + A T^{-1})^{-1}$ exists if

$$\operatorname{spr} A < (\operatorname{spr} T^{-1})^{-1} . \tag{4.26}$$

5. Operator-valued functions

Operator-valued functions $T_t = T(t)$ defined for a real or complex variable t and taking values in $\mathscr{B}(X, Y)$ can be defined and treated just as vector-valued functions $u(t)$ were in § 1.7. A new feature for $T(t)$ appears again since the products $T(t)\, u(t)$ and $S(t)\, T(t)$ are defined. Thus we have, for example, the formulas

$$\begin{aligned} \frac{d}{dt} T(t)\, u(t) &= T'(t)\, u(t) + T(t)\, u'(t) , \\ \frac{d}{dt} T(t)\, S(t) &= T'(t)\, S(t) + T(t)\, S'(t) , \end{aligned} \tag{4.27}$$

whenever the products are meaningful and the derivatives on the right exist. Also we have

$$\frac{d}{dt} T(t)^{-1} = - T(t)^{-1}\, T'(t)\, T(t)^{-1} \tag{4.28}$$

whenever $T(t)^{-1}$ and $T'(t)$ exist. This follows from the identity

$$S^{-1} - T^{-1} = - S^{-1}(S - T)\, T^{-1} \tag{4.29}$$

and the continuity of T^{-1} as function of T proved in par. 4.

For the integrals of operator-valued functions, we have formulas similar to (1.30). In addition, we have

$$\begin{aligned} &\textstyle\int S u(t)\, dt = S \int u(t)\, dt , \quad \int T(t)\, u\, dt = (\int T(t)\, dt)\, u , \\ &\textstyle\int S T(t)\, dt = S \int T(t)\, dt , \quad \int T(t)\, S\, dt = (\int T(t)\, dt)\, S . \end{aligned} \tag{4.30}$$

Of particular importance again are *holomorphic* functions $T(t)$ of a complex variable t; here the same remarks apply as those given for vector-valued functions (see § 1.7). It should be added that $S(t)\, T(t)$ and $T(t)\, u(t)$ are holomorphic if all factors are holomorphic, and that $T(t)^{-1}$ is holomorphic whenever $T(t)$ is holomorphic and $T(t)^{-1}$ exists [the latter follows from (4.28)].

Naturally there are restrictions in applying complex function theory to vector- and operator-valued functions. Multiplication is not defined for two vector-valued functions. It is defined for operator-valued functions, but division is not always possible. Thus $T(t)^{-1}$ need not exist for a holomorphic operator-valued function $T(t)$. It will be shown below (see Theorem II-1.5a), however, that $T(t)^{-1}$ exists as a *meromorphic* function if $T(t)$ is holomorphic (or even meromorphic) and if $T(t_0)^{-1}$ exists at a single point t_0.

Example 4.8. The exponential function e^{tT} defined by (4.20) is an *entire function* of t (holomorphic in the whole complex plane), with

$$\frac{d}{dt} e^{tT} = T e^{tT} = e^{tT} T . \tag{4.31}$$

Example 4.9. Consider the Neumann series

$$S(t) = (1 - tT)^{-1} = \sum_{n=0}^{\infty} t^n T^n \tag{4.32}$$

with a complex parameter t. By Problem 4.6 this series is absolutely convergent for $|t| < 1/\operatorname{spr} T$. Actually, *the convergence radius r of* (4.32) *is exactly equal to* $1/\operatorname{spr} T$. Since $S(t)$ is holomorphic for $|t| < r$, the Cauchy inequality gives $\|T^n\| \leqq M_{r'} r'^{-n}$ for all n and $r' < r$ as in the case of numerical power series[1] ($M_{r'}$ is independent of n). Hence $\operatorname{spr} T = \lim \|T^n\|^{1/n} \leqq r'^{-1}$ and, going to the limit $r' \to r$, we have $\operatorname{spr} T \leqq r^{-1}$ or $r \leqq 1/\operatorname{spr} T$. Since the opposite inequality was proved above, this gives the proof of the required result. Incidentally, it follows that $\operatorname{spr} T$ is independent of the norm used in its definition.

6. Pairs of projections

As an application of analysis with operators and also with a view to later applications, we shall prove some theorems concerning a pair of projections (idempotents)[2]. As defined in § 3.4, a projection P is an operator of $\mathscr{B}(\mathsf{X})$ such that $P^2 = P$. $1 - P$ is a projection with P.

Let $P, Q \in \mathscr{B}(\mathsf{X})$ be two projections. Then

$$R = (P - Q)^2 = P + Q - PQ - QP \tag{4.33}$$

commutes with P and Q; this is seen by noting that $PR = P - PQP = RP$ and similarly for Q. For the same reason $(1 - P - Q)^2$ commutes with P and Q because $1 - P$ is a projection. Actually we have the identity

$$(P - Q)^2 + (1 - P - Q)^2 = 1 \tag{4.34}$$

as is verified by direct computation. Another useful identity is

[1] We have $T^n = (2\pi i)^{-1} \int_{|t| = r'} t^{-n-1} S(t)\, dt$ and so $\|T^n\| \leqq (2\pi)^{-1} \int_{|t| = r'} r'^{-n-1} \cdot \|S(t)\|\, |dt| \leqq r'^{-n} M_{r'}$, where $M_{r'} = \max_{|t| = r'} \|S(t)\| < \infty$.

[2] The following results, which are taken from T. Kato [9], are true even when X is an ∞-dimensional Banach space. For the special case of projections in a unitary (Hilbert) space, see § 6.8. For related results cf. Akhiezer and Glazman [[1]], Sz.-Nagy [1], [2], Wolf [1].

(4.35) $$(PQ - QP)^2 = (P - Q)^4 - (P - Q)^2 = R^2 - R\,,$$

the proof of which is again straightforward and will be left to the reader.

Set

(4.36) $$U' = QP + (1 - Q)(1 - P)\,, \quad V' = PQ + (1 - P)(1 - Q)\,.$$

U' maps $\mathsf{R}(P) = P\mathsf{X}$ into $Q\mathsf{X}$ and $(1 - P)\mathsf{X}$ into $(1 - Q)\mathsf{X}$, whereas V' maps $Q\mathsf{X}$ into $P\mathsf{X}$ and $(1 - Q)\mathsf{X}$ into $(1 - P)\mathsf{X}$. But these mappings are not inverse to each other; in fact it is easily seen that

(4.37) $$V'U' = U'V' = 1 - R\,.$$

A pair of mutually inverse operators U, V with the mapping properties stated above can be constructed easily, however, since R commutes with P, Q and therefore with U', V' too. It suffices to set

(4.38) $$\begin{aligned} U &= U'(1 - R)^{-1/2} = (1 - R)^{-1/2}\,U'\,, \\ V &= V'(1 - R)^{-1/2} = (1 - R)^{-1/2}\,V'\,, \end{aligned}$$

provided the inverse square root $(1 - R)^{-1/2}$ of $1 - R$ exists. A natural definition of this operator is given by the binomial series

(4.39) $$(1 - R)^{-1/2} = \sum_{n=0}^{\infty} \binom{-1/2}{n} (-R)^n\,.$$

This series is absolutely convergent if $\|R\| < 1$ or, more generally, if

(4.40) $$\operatorname{spr} R < 1\,,$$

and the sum T of this series satisfies the relation $T^2 = (1 - R)^{-1}$ just as in the numerical binomial series. Thus[1]

(4.41) $$VU = UV = 1\,, \quad V = U^{-1}\,, \quad U = V^{-1}\,.$$

Since $U'P = QP = QU'$ and $PV' = PQ = V'Q$ as is seen from (4.36), we have $UP = QU$ and $PV = VQ$ by the commutativity of R with all other operators here considered. Thus we have

(4.42) $$Q = UPU^{-1}\,, \quad P = U^{-1}QU\,.$$

Thus P and Q are *similar* to each other (see § 5.7). They are isomorphic to each other in the sense that any linear relationship such as $v = Pu$ goes over to $v' = Qu'$ by the one-to-one linear mapping $U: u' = Uu$, $v' = Uv$. In particular their ranges $P\mathsf{X}$, $Q\mathsf{X}$ are *isomorphic*, being mapped onto each other by U and U^{-1}. Thus

(4.43) $$\dim P = \dim Q\,, \quad \dim(1 - P) = \dim(1 - Q)\,.$$

[1] As is shown later (§ 6.7), U and V are unitary if X is a unitary space and P, Q are orthogonal projections. The same U appears also in Sz.-Nagy [1] and Wolf [1] with a different expression; its identity with the above U was shown in T. Kato [9].

An immediate consequence of this result is

Lemma 4.10. *Let $P(t)$ be a projection depending continuously on a parameter t varying in a (connected) region of real or complex numbers. Then the ranges $P(t)\,\mathsf{X}$ for different t are isomorphic to one another. In particular* $\dim P(t)\,\mathsf{X}$ *is constant.*

To see this, it suffices to note that $\|P(t') - P(t'')\| < 1$ for sufficiently small $|t' - t''|$ so that the above result applies to the pair $P(t'), P(t'')$.

Problem 4.11. Under the assumption (4.40), we have $PQ\mathsf{X} = P\mathsf{X}$, $QP\mathsf{X} = Q\mathsf{X}$ [hint: $PQ = PV' = PU^{-1}(1 - R)^{1/2}$].

Problem 4.11a. Similarity of P with Q is also implemented by $Z = (1 - R)^{-1/2}(1 - P - Q)$ if $\operatorname{spr} R < 1 : Q = ZPZ^{-1}$. Moreover, $Z = U(1 - 2P) = (1 - 2Q)\,U$.

Problem 4.12. For any two projections P, Q, we have

$$(4.44)\qquad (1 - P + QP)\,(1 - Q + PQ) = 1 - R \quad [R \text{ is given by (4.33)}]\,,$$

$$(4.45)\qquad (1 - P + QP)^{-1} = (1 - R)^{-1}(1 - Q + PQ) \quad \text{if} \quad \operatorname{spr} R < 1\,.$$

If $\operatorname{spr} R < 1$, $W = 1 - P + QP$ maps $P\mathsf{X}$ onto $Q\mathsf{X}$ and $Wu = u$ for $u \in (1 - P)\,\mathsf{X}$, while W^{-1} maps $Q\mathsf{X}$ onto $P\mathsf{X}$ and $W^{-1}u = u$ for $u \in (1 - P)\,\mathsf{X}$, and we have $\mathsf{X} = Q\mathsf{X} \oplus (1 - P)\,\mathsf{X}$.

Problem 4.13. For any two projections P, Q such that $\operatorname{spr}(P - Q)^2 < 1$, there is a family $P(t)$, $0 \leqq t \leqq 1$, of projections depending holomorphically on t such that $P(0) = P$, $P(1) = Q$. [hint: set $2P(t) = 1 + (2P - 1 + 2t(Q - P)) \cdot (1 - 4t(1 - t)\,R)^{-1/2}$.]

7. Product formulas

Lemma 4.14. *Let $A \in \mathscr{B}(\mathsf{X})$. Let $B_n(t) \in \mathscr{B}(\mathsf{X})$, $n = 1, 2, \ldots$, with $\|B_n(t)\| \to 0$ as $n \to \infty$, uniformly in $t \in \mathrm{D}$, where D is a bounded region in the complex plane. Then*

$$(4.46)\qquad \lim_{n\to\infty} [1 + (t/n)\,(A + B_n(t))]^n = e^{tA}$$

uniformly for $t \in \mathrm{D}$.

Proof. Write $V_n(t) = 1 + (t/n)\,(A + B_n(t))$. Then

$$(4.47)\qquad \begin{aligned} \|V_n^k\| &\leqq [1 + |t/n|\,(\|A\| + \|B_n\|)]^k \\ &\leqq [1 + |t/n|\,(\|A\| + \|B_n\|)]^n \leqq e^{|t|\,(\|A\| + \|B_n\|)} \leqq M \end{aligned}$$

for $0 \leqq k \leqq n$ and $t \in \mathrm{D}$, where M is a constant. Here we have suppressed the variable t in $B_n(t)$ and $V_n(t)$.

If we write $U_n(t) = e^{(t/n)A}$, we have $U_n^n = e^{tA}$ and $\|U_n^k\| \leqq M$, too. Moreover,

$$(4.48)\qquad V_n^n - U_n^n = \sum_{j=0}^{n-1} V_n^j\,(V_n - U_n)\,U_n^{n-j-1}\,.$$

Since

$$\begin{aligned} V_n - U_n &= 1 + (t/n)\,(A + B_n) - [1 + (t/n)\,A + (t^2/2n^2)\,A^2 + \cdots] \\ &= (t/n)\,[B_n - (t/2n)\,A^2 + \cdots]\,, \end{aligned}$$

we obtain

$$\|V_n - U_n\| \leqq |t/n| \, \|B_n\| + |t/n|^2 \, \| A \|^2 \tag{4.49}$$

if n is sufficiently large.

It follows from these estimates that

$$\| V_n^n - U_n^n\| \leqq n M^2 \, [|t/n| \, \| B_n\| + |t/n|^2 \, \| A \|^2] \to 0 \quad \text{as} \quad n \to \infty, \tag{4.50}$$

uniformly in $t \in \mathsf{D}$. This proves the lemma (recall that $U_n^n = e^{tA}$).

Corollary 4.15. *For any $A \in \mathscr{B}(\mathsf{X})$, one has*

$$e^{tA} = \lim_{n\to\infty} [1 + (t/n) \, A]^n = \lim_{n\to\infty} [1 - (t/n) \, A]^{-n} \tag{4.51}$$

uniformly in any bounded region of the complex parameter t.

Corollary 4.16. (*The Lie-Trotter product formula*). *For any $A, B \in \mathscr{B}(\mathsf{X})$, one has*

$$\lim_{n\to\infty} [e^{(t/n)A} \, e^{(t/n)B}]^n = e^{t(A+B)} \tag{4.52}$$

uniformly in any bounded region of the complex parameter t.

Proof. The first formula in (4.51) follows from Lemma 4.14 by setting $B_n(t) = 0$. The second formula follows from

$$[1 - (t/n) \, A]^{-1} = 1 + (t/n) \, A + o\,(1/n) \, ,$$

where $o\,(1/n)$ denotes an operator whose norm is $o\,(1/n)$ as $n \to \infty$, uniformly in t in any bounded region. Similarly, (4.52) follows from

$$\begin{aligned} e^{(t/n)A} \, e^{(t/n)B} &= [1 + (t/n) \, A + o\,(1/n)] \, [1 + (t/n) \, B + o\,(1/n)] \\ &= 1 + (t/n) \, (A + B) + o\,(1/n) \, . \end{aligned}$$

Remark 4.17. In (4.52), the factor $e^{(t/n)A}$ may be replaced by $1 + (t/n) \, A$, or $(1 - (t/n) \, A)^{-1}$, or $(1 - (t/n) \, (A/k))^{-k}$, or, more generally, by $\phi\,(tA/n)$, where $\phi\,(\zeta)$ is any function holomorphic at $\zeta = 0$ with $\phi\,(0) = \phi'\,(0) = 1$. Similarly for the factor $e^{(t/n)B}$. Moreover, there may be any number of such factors $\phi_j\,(tA_j/n)$ $(j = 1, 2, \ldots, r)$ instead of two, the limit being equal to $\exp\,[t\,(A_1 + \cdots + A_r)]$.

§ 5. The eigenvalue problem

1. Definitions

In this section X denotes a given vector space with $0 < \dim \mathsf{X} = N < \infty$, but we shall consider X a normed space whenever convenient by introducing an appropriate norm.

Let $T \in \mathscr{B}(\mathsf{X})$. A complex number λ is called an *eigenvalue (proper value, characteristic value)* of T if there is a non-zero vector $u \in \mathsf{X}$ such that

$$T u = \lambda \, u \, . \tag{5.1}$$

u is called an *eigenvector (proper vector, characteristic vector)* of T belonging to (associated with, etc.) the eigenvalue λ. The set N_λ of all $u \in \mathsf{X}$ such that $Tu = \lambda u$ is a linear manifold of X; it is called the *(geometric) eigenspace* of T for the eigenvalue λ, and dim N_λ is called the *(geometric) multiplicity* of λ. N_λ is defined even when λ is not an eigenvalue; then we have $\mathsf{N}_\lambda = 0$. In this case it is often convenient to say that N_λ is the eigenspace for the eigenvalue λ with multiplicity zero, though this is not in strict accordance with the definition of an eigenvalue.

Problem 5.1. λ is an eigenvalue of T if and only if $\lambda - \zeta$ is an eigenvalue of $T - \zeta$. N_λ is the null space of $T - \lambda$, and the geometric multiplicity of the eigenvalue λ of T is the nullity of $T - \lambda$. $T - \lambda$ is singular if and only if λ is an eigenvalue of T.

It can easily be proved that *eigenvectors of T belonging to different eigenvalues are linearly independent*. It follows that *there are at most N eigenvalues of T*. The set of all eigenvalues of T is called the *spectrum* of T; we denote it by $\Sigma(T)$. Thus $\Sigma(T)$ is a finite set with not more than N points.

The *eigenvalue problem* consists primarily in finding all eigenvalues and eigenvectors (or eigenspaces) of a given operator T. A vector $u \neq 0$ is an eigenvector of T if and only if the one-dimensional linear manifold $[u]$ spanned by u is invariant under T (see § 3.5). Thus the eigenvalue problem is a special case of the problem of determining all invariant linear manifolds for T (a generalized form of the eigenvalue problem).

If M is an invariant subspace of T, the part T_M of T in M is defined. As is easily seen, any eigenvalue [eigenvector] of T_M is an eigenvalue [eigenvector] for T. For convenience an eigenvalue of T_M is called an *eigenvalue of T in* M.

If there is a projection P that commutes with T, T is decomposed according to the decomposition $\mathsf{X} = \mathsf{M} \oplus \mathsf{N}$, $\mathsf{M} = P\mathsf{X}$, $\mathsf{N} = (1 - P)\,\mathsf{X}$ (see § 3.5). To solve the eigenvalue problem for T, it is then sufficient to consider the eigenvalue problem for the parts of T in M and in N[1]. The part T_M of T in M may be identified with the operator $PT = TP = PTP$ in the sense stated in Remark 3.15. It should be noticed, however, that TP has an eigenvalue zero with the eigenspace N and therefore with multiplicity $N - m$ (where $m = \dim P\mathsf{X}$) in addition to the eigenvalues of T_M[2].

Problem 5.2. No eigenvalue of T exceeds $\|T\|$ in absolute value, where $\| \ \|$ is the norm of $\mathscr{B}(\mathsf{X})$ associated with any norm of X.

---

[1] If $Tu = \lambda u$, then $TPu = PTu = \lambda Pu$ so that $Pu \in \mathsf{M}$ is, if not 0, an eigenvector of T (and of T_M) for the eigenvalue λ, and similarly for $(1 - P)\,u$. Thus any eigenvalue of T must be an eigenvalue of at least one of T_M and T_N, and any eigenvector of T is the sum of eigenvectors of T_M and T_N for the same eigenvalue. The eigenspace of T for λ is the direct sum of the eigenspaces of T_M and T_N for λ.

[2] Strictly speaking, this is true only when T_M has no eigenvalue 0. If T_M has the eigenvalue 0 with the eigenspace L, the eigenspace for the eigenvalue 0 of TP is $\mathsf{N} \oplus \mathsf{L}$.

Problem 5.3. If T has a diagonal matrix with respect to a basis, the eigenvalues of T coincide with (the different ones among) all its diagonal elements.

2. The resolvent

Let $T \in \mathscr{B}(\mathsf{X})$ and consider the inhomogeneous linear equation

$$(T - \zeta)\, u = v\,, \tag{5.2}$$

where ζ is a given complex number, $v \in \mathsf{X}$ is given and $u \in \mathsf{X}$ is to be found. In order that this equation have a solution u for every v, it is necessary and sufficient that $T - \zeta$ be nonsingular, that is, ζ be different from any eigenvalue λ_h of T. Then the inverse $(T - \zeta)^{-1}$ exists and the solution u is given by

$$u = (T - \zeta)^{-1}\, v\,. \tag{5.3}$$

The operator-valued function

$$R(\zeta) = R(\zeta, T) = (T - \zeta)^{-1} \tag{5.4}$$

is called the *resolvent*[1] of T. The complementary set of the spectrum $\Sigma(T)$ (that is, the set of all complex numbers different from any of the eigenvalues of T) is called the *resolvent set* of T and will be denoted by $\mathrm{P}(T)$. The resolvent $R(\zeta)$ is thus defined for $\zeta \in \mathrm{P}(T)$.

Problem 5.4. $R(\zeta)$ commutes with T. $R(\zeta)$ has exactly the eigenvalues $(\lambda_h - \zeta)^{-1}$.

Problem 5.4a. If $T \in \mathscr{B}(\mathsf{X}, \mathsf{Y})$ and $S = \mathscr{B}(\mathsf{Y}, \mathsf{X})$, then $S\,T \in \mathscr{B}(\mathsf{X})$ and $TS \in \mathscr{B}(\mathsf{Y})$ have the same spectrum except for possible eigenvalue zero, and $\zeta\, R(\zeta, ST) = -1_{\mathsf{X}} + S\, R(\zeta, TS)\, T$. In particular, $\Sigma(ST) = \Sigma(TS)$ if $\mathsf{Y} = \mathsf{X}$. Here 1_{X} denotes the identity operator in X.

An important property of the resolvent is that it satisfies the (first) *resolvent equation*

$$R(\zeta_1) - R(\zeta_2) = (\zeta_1 - \zeta_2)\, R(\zeta_1)\, R(\zeta_2)\,. \tag{5.5}$$

This is seen easily if one notes that the left member is equal to $R(\zeta_1) \cdot \cdot (T - \zeta_2)\, R(\zeta_2) - R(\zeta_1)\, (T - \zeta_1)\, R(\zeta_2)$. In particular, (5.5) implies that *$R(\zeta_1)$ and $R(\zeta_2)$ commute.* Also we have $R(\zeta_1) = [1 - (\zeta_2 - \zeta_1)\, R(\zeta_1)] \cdot \cdot R(\zeta_2)$. According to the results on the Neumann series (see Example 4.5), this leads to the expansion

$$R(\zeta) = [1 - (\zeta - \zeta_0)\, R(\zeta_0)]^{-1} R(\zeta_0) = \sum_{n=0}^{\infty} (\zeta - \zeta_0)^n\, R(\zeta_0)^{n+1}\,, \tag{5.6}$$

the series being absolutely convergent at least if

$$|\zeta - \zeta_0| < \|R(\zeta_0)\|^{-1} \tag{5.7}$$

[1] The resolvent is the operator-valued function $\zeta \to R(\zeta)$. It appears, however, that also the value $R(\zeta)$ of this function at a particular ζ is customarily called the resolvent. Sometimes $(\zeta - T)^{-1}$ instead of $(T - \zeta)^{-1}$ is called the resolvent. In this book we follow the definition of Stone [[1]].

for *some* norm. We shall refer to (5.6) as the *first Neumann series for the resolvent.*

(5.6) shows that $R(\zeta)$ is holomorphic in ζ with the Taylor series[1] shown on the right; hence

$$\left(\frac{d}{d\zeta}\right)^n R(\zeta) = n!\, R(\zeta)^{n+1}, \quad n = 1, 2, 3, \ldots . \tag{5.8}$$

According to Example 4.9, the convergence radius of the series of (5.6) is equal to $1/\operatorname{spr} R(\zeta_0)$. Hence this series is convergent if and only if

$$|\zeta - \zeta_0| < 1/\operatorname{spr} R(\zeta_0) = (\lim \|R(\zeta_0)^n\|^{1/n})^{-1} . \tag{5.9}$$

For large $|\zeta|$, $R(\zeta)$ has the expansion

$$R(\zeta) = -\zeta^{-1}(1 - \zeta^{-1} T)^{-1} = - \sum_{n=0}^{\infty} \zeta^{-n-1} T^n , \tag{5.10}$$

which is convergent if and only if $|\zeta| > \operatorname{spr} T$; thus $R(\zeta)$ is holomorphic at infinity.

Problem 5.5. We have

$$\|R(\zeta)\| \leqq (|\zeta| - \|T\|)^{-1}, \quad \|R(\zeta) + \zeta^{-1}\| \leqq |\zeta|^{-1}(|\zeta| - \|T\|)^{-1}\|T\| \quad \text{for} \quad |\zeta| > \|T\| . \tag{5.11}$$

The spectrum $\Sigma(T)$ is never empty; T has at least one eigenvalue. Otherwise $R(\zeta)$ would be an entire function such that $R(\zeta) \to 0$ for $\zeta \to \infty$ [see (5.11)]; then we must have $R(\zeta) = 0$ by Liouville's theorem[2]. But this gives the contradiction $1 = (T - \zeta)\, R(\zeta) = 0$[3].

It is easily seen that each eigenvalue of T is a singularity[4] of the analytic function $R(\zeta)$. Since there is at least one singularity of $R(\zeta)$ on the convergence circle[5] $|\zeta| = \operatorname{spr} T$ of (5.10), $\operatorname{spr} T$ coincides with the largest (in absolute value) eigenvalue of T:

$$\operatorname{spr} T = \max_h |\lambda_h| . \tag{5.12}$$

This shows again that $\operatorname{spr} T$ is independent of the norm used in its definition.

Problem 5.6. $\operatorname{spr} T = 0$ if and only if T is nilpotent. [hint for "only if" part: If $\operatorname{spr} T = 0$, we have also $\operatorname{spr} T_{\mathsf{M}} = 0$ for any part T_{M} of T in an invariant subspace M, so that T_{M} is singular. Thus the part of T in each of the invariant sub-

---

[1] This is an example of the use of function theory for operator-valued functions; see Knopp [[1]], p. 79.

[2] See Knopp [[1]], p. 113. Liouville's theorem implies that $R(\zeta)$ is constant; since $R(\zeta) \to 0$ for $\zeta \to \infty$, this constant is 0.

[3] Note that we assume $\dim \mathsf{X} > 0$.

[4] Suppose λ is a regular point (removable singularity) of the analytic function $R(\zeta)$. Then $\lim_{\zeta \to \lambda} R(\zeta) = R$ exists and so $(T - \lambda) R = \lim_{\zeta \to \lambda} (T - \zeta) R(\zeta) = 1$. Thus $(T - \lambda)^{-1} = R$ exists and λ is not an eigenvalue.

[5] See Knopp [[1]], p. 101.

spaces $T^n\mathsf{X}$ (see Problem 3.12) is singular. Hence $\mathsf{X}\supset T\mathsf{X}\supset T^2\mathsf{X}\supset\cdots$ with all inclusions $\supset$ being proper until 0 is reached.]

3. Singularities of the resolvent

The singularities of $R(\zeta)$ are exactly the eigenvalues λ_h, $h = 1, \ldots, s$, of T. Let us consider the Laurent series[1] of $R(\zeta)$ at $\zeta = \lambda_h$. For simplicity we may assume for the moment that $\lambda_h = 0$ and write

$$R(\zeta) = \sum_{n=-\infty}^{\infty} \zeta^n A_n\,. \tag{5.13}$$

The coefficients A_n are given by

$$A_n = \frac{1}{2\pi i}\int_\Gamma \zeta^{-n-1} R(\zeta)\, d\zeta\,, \tag{5.14}$$

where Γ is a positively-oriented small circle enclosing $\zeta = 0$ but excluding other eigenvalues of T. Since Γ may be expanded to a slightly larger circle Γ' without changing (5.14), we have

$$\begin{aligned} A_n A_m &= \left(\frac{1}{2\pi i}\right)^2 \int_{\Gamma'}\int_{\Gamma} \zeta^{-n-1}\zeta'^{-m-1} R(\zeta)\, R(\zeta')\, d\zeta\, d\zeta' \\ &= \left(\frac{1}{2\pi i}\right)^2 \int_{\Gamma'}\int_{\Gamma} \zeta^{-n-1}\zeta'^{-m-1} (\zeta'-\zeta)^{-1}[R(\zeta') - R(\zeta)]\, d\zeta\, d\zeta'\,, \end{aligned}$$

where the resolvent equation (5.5) has been used. The double integral on the right may be computed in any order. Considering that Γ' lies outside Γ, we have

$$\begin{aligned} \frac{1}{2\pi i}\int_\Gamma \zeta^{-n-1}(\zeta'-\zeta)^{-1}\, d\zeta &= \eta_n\, \zeta'^{-n-1} \\ \frac{1}{2\pi i}\int_{\Gamma'} \zeta'^{-m-1}(\zeta'-\zeta)^{-1}\, d\zeta' &= (1-\eta_m)\, \zeta^{-m-1} \end{aligned} \tag{5.15}$$

where the symbol η_n is defined by

$$\eta_n = 1 \quad \text{for} \quad n \geqq 0 \quad \text{and} \quad \eta_n = 0 \quad \text{for} \quad n < 0\,. \tag{5.16}$$

Thus

$$A_n A_m = \frac{\eta_n + \eta_m - 1}{2\pi i}\int_\Gamma \zeta^{-n-m-2} R(\zeta)\, d\zeta = (\eta_n + \eta_m - 1)\, A_{n+m+1}\,. \tag{5.17}$$

For $n = m = -1$ this gives $A_{-1}^2 = -A_{-1}$. Thus $-A_{-1}$ is a projection, which we shall denote by P. For $n, m < 0$, (5.17) gives $A_{-2}^2 = -A_{-3}$, $A_{-2}A_{-3} = -A_{-4}$, ... On setting $-A_{-2} = D$, we thus obtain $A_{-k} = -D^{k-1}$ for $k \geqq 2$. Similarly we obtain $A_n = S^{n+1}$ for $n \geqq 0$ with $S = A_0$.

[1] See KNOPP [[1]], p. 117.

Returning to the general case in which $\zeta = \lambda_h$ is the singularity instead of $\zeta = 0$, we see that the Laurent series takes the form

$$R(\zeta) = -(\zeta - \lambda_h)^{-1} P_h - \sum_{n=1}^{\infty} (\zeta - \lambda_h)^{-n-1} D_h^n + \\ + \sum_{n=0}^{\infty} (\zeta - \lambda_h)^n S_h^{n+1} . \tag{5.18}$$

Setting in (5.17) $n = -1, m = -2$ and then $n = -1, m = 0$, we see that

$$P_h D_h = D_h P_h = D_h , \quad P_h S_h = S_h P_h = 0 . \tag{5.19}$$

Thus the two lines on the right of (5.18) represent a decomposition of the operator $R(\zeta)$ according to the decomposition $\mathsf{X} = \mathsf{M}_h \oplus \mathsf{M}_h'$, where $\mathsf{M}_h = P_h \mathsf{X}$ and $\mathsf{M}_h' = (1 - P_h)\mathsf{X}$. As the principal part of a Laurent series at an isolated singularity, the first line of (5.18) is convergent for $\zeta - \lambda_h \neq 0$, so that the part of $R(\zeta)$ in M_h has only the one singularity $\zeta = \lambda_h$, and the spectral radius of D_h must be zero. According to Problem 5.6, it follows that D_h is nilpotent and therefore (see Problem 3.10)

$$D_h^{m_h} = 0 , \quad m_h = \dim \mathsf{M}_h = \dim P_h . \tag{5.20}$$

Thus the principal part in the Laurent expansion (5.18) of $R(\zeta)$ is finite, $\zeta = \lambda_h$ being a *pole* of order not exceeding m_h. Since the same is true of all singularities λ_h of $R(\zeta)$, $R(\zeta)$ is a *meromorphic function*[1].

The P_h for different h satisfy the following relations:

$$P_h P_k = \delta_{hk} P_h , \quad \sum_{h=1}^{s} P_h = 1 , \quad P_h T = T P_h . \tag{5.21}$$

The first relation can be proved in the same way as we proved $P_h^2 = P_h$ above [which is a special case of (5.17)]; it suffices to notice that

$$P_h = -\frac{1}{2\pi i} \int_{\Gamma_h} R(\zeta)\, d\zeta \tag{5.22}$$

where the circles Γ_h for different h lie outside each other. The second equality of (5.21) is obtained by integrating $R(\zeta)$ along a large circle enclosing all the eigenvalues of T and noting the expansion (5.10) of $R(\zeta)$ at infinity. The commutativity of P_h with T follows immediately from (5.22).

Since $R(\zeta)$ is meromorphic and regular at infinity, the decomposition of $R(\zeta)$ into *partial fractions*[2] takes the form

$$R(\zeta) = -\sum_{h=1}^{s} \left[(\zeta - \lambda_h)^{-1} P_h + \sum_{n=1}^{m_h - 1} (\zeta - \lambda_h)^{-n-1} D_h^n \right] . \tag{5.23}$$

Problem 5.7. $\operatorname{spr} R(\zeta) = \left[\min_h |\zeta - \lambda_h|\right]^{-1} = [\operatorname{dist}(\zeta, \Sigma(T))]^{-1}$.

Problem 5.8. We have

$$P_h D_k = D_k P_h = \delta_{hk} D_h ; \quad D_h D_k = 0 , \quad h \neq k . \tag{5.24}$$

[1] See KNOPP [[2]], p. 34.

[2] See KNOPP [[2]], p. 34.

Problem 5.9. For any simple closed (rectifiable) curve Γ with positive direction and not passing through any eigenvalue λ_h, we have

$$\frac{1}{2\pi i}\int_\Gamma R(\zeta)\, d\zeta = -\sum P_h, \tag{5.25}$$

where the sum is taken for those h for which λ_h is inside Γ.

Multiplying (5.14) by T from the left or from the right and noting that $TR(\zeta) = R(\zeta)T = 1 + \zeta R(\zeta)$, we obtain $TA_n = A_n T = \delta_{n0} + A_{n-1}$. If the singularity is at $\zeta = \lambda_h$ instead of $\zeta = 0$, this gives for $n = 0$ and $n = -1$

$$\begin{aligned}(T-\lambda_h)\, S_h &= S_h(T-\lambda_h) = 1 - P_h\,,\\ P_h(T-\lambda_h) &= (T-\lambda_h)\, P_h = D_h\,.\end{aligned} \tag{5.26}$$

For each $h = 1, \ldots, s$, the holomorphic part in the Laurent expansion (5.18) will be called the *reduced resolvent* of T with respect to the eigenvalue λ_h; we denote it by $S_h(\zeta)$:

$$S_h(\zeta) = \sum_{n=0}^{\infty} (\zeta-\lambda_h)^n\, S_h^{n+1}\,. \tag{5.27}$$

It follows from (5.19) and (5.26) that

$$S_h = S_h(\lambda_h)\,, \quad S_h(\zeta)\, P_h = P_h\, S_h(\zeta) = 0\,, \tag{5.28}$$

$$(T-\zeta)\, S_h(\zeta) = S_h(\zeta)\,(T-\zeta) = 1 - P_h\,. \tag{5.29}$$

The last equalities show that the parts of $T - \zeta$ and of $S_h(\zeta)$ in the invariant subspace $\mathsf{M}_h' = (1 - P_h)\mathsf{X}$ are inverse to each other.

Problem 5.10. We have

$$(T-\lambda_h)\, D_h^n = D_h^{n+1}\,, \quad n = 1, 2, \ldots. \tag{5.30}$$

$$(T-\lambda_h)^{m_h}\, P_h = 0\,. \tag{5.31}$$

$$S_h(\zeta) = -\sum_{k\neq h}\left[(\zeta-\lambda_k)^{-1} P_k + \sum_{n=1}^{m_k-1}(\zeta-\lambda_k)^{-n-1} D_k^n\right]. \tag{5.32}$$

Problem 5.11. Each $S_h(\zeta)$ satisfies the resolvent equation [see (5.5)] and

$$\left(\frac{d}{d\zeta}\right)^n S_h(\zeta) = n!\, S_h(\zeta)^{n+1}\,, \quad n = 1, 2, \ldots. \tag{5.33}$$

4. The canonical form of an operator

The result of the preceding paragraph leads to the *canonical form* of the operator T. Denoting as above by M_h the range of the projection P_h, $h = 1, \ldots, s$, we have

$$\mathsf{X} = \mathsf{M}_1 \oplus \cdots \oplus \mathsf{M}_s\,. \tag{5.34}$$

Since the P_h commute with T and with one another, the M_h are invariant subspaces for T and T is decomposed according to the decomposition (5.34) (see § 3.5). M_h is called the *algebraic eigenspace* (or *principal subspace*) for the eigenvalue λ_h of T, and $m_h = \dim \mathsf{M}_h$ is the *algebraic multiplicity* of λ_h. In what follows P_h will be called the *eigenprojection* and D_h the *eigennilpotent* for the eigenvalue λ_h of T. Any vector $u \neq 0$

of M_h is called a *generalized eigenvector* (or *principal vector*) for the eigenvalue λ_h of T.

It follows from (5.26) that

$$(5.35) \qquad T P_h = P_h T = P_h T P_h = \lambda_h P_h + D_h , \quad h = 1, \ldots, s .$$

Thus the part T_{M_h} of T in the invariant subspace M_h is the sum of the scalar operator λ_h and a nilpotent D_{h,M_h}, the part of D_h in M_h. As is easily seen, T_{M_h} has one and only one eigenvalue λ_h.

Addition of the s equations (5.35) gives by (5.21)

$$(5.36) \qquad T = S + D$$

where

$$(5.37) \qquad S = \sum_h \lambda_h P_h$$

$$(5.38) \qquad D = \sum_h D_h .$$

An operator S of the form (5.37) where $\lambda_h \neq \lambda_k$ for $h \neq k$ and the P_h satisfy (3.27–3.28) is said to be *diagonalizable* (or *diagonable* or *semisimple*). S is the direct sum (see § 3.5) of scalar operators. D is nilpotent, for it is the direct sum of nilpotents D_h and so $D^n = \sum D_h^n = 0$ for $n \geqq \max m_h$. It follows from (5.24) that D commutes with S.

(5.36) shows that *every operator $T \in \mathscr{B}(\mathsf{X})$ can be expressed as the sum of a diagonable operator S and a nilpotent D that commutes with S.*

The eigenvalue λ_h of T will also be said to be *semisimple* if the associated eigennilpotent D_h is zero, and *simple*[1] if $m_h = 1$ (note that $m_h = 1$ implies $D_h = 0$). T is diagonable if and only if all its eigenvalues are semisimple. T is said to be *simple* if all the eigenvalues λ_h are simple; in this case T has N eigenvalues.

(5.36) is called the *spectral representation* of T. The spectral representation is unique in the following sense: if T is the sum of a diagonable operator S and a nilpotent D that commutes with S, then S and D must be given by (5.37) and (5.38) respectively. To show this, we first note that any operator R that commutes with a diagonable operator S of the form (5.37) commutes with every P_h so that M_h is invariant under R. In fact, multiplying $RS = SR$ from the left by P_h and from the right by P_k, we have by $P_h P_k = \delta_{hk}$

$$\lambda_k P_h R P_k = \lambda_h P_h R P_k \quad \text{or} \quad P_h R P_k = 0 \quad \text{for} \quad h \neq k .$$

Addition of the results for all $k \neq h$ for fixed h gives $P_h R (1 - P_h) = 0$ or $P_h R = P_h R P_h$. Similarly we obtain $R P_h = P_h R P_h$ and therefore $R P_h = P_h R$.

Suppose now that

$$T = S' + D' , \quad S' = \sum_{h=1}^{s'} \lambda_h' P_h' , \quad D' S' = S' D' ,$$

[1] An eigenvalue which is not simple is said to be *degenerate*.

is a second expression of T in the form (5.36). By what was just proved, we have $D' = \sum D'_h$ with $D'_h = P'_h D' = D' P'_h$. Hence $T - \zeta = \sum [(\lambda'_h - \zeta) P'_h + D'_h]$ and therefore

$$(5.39) \qquad (T-\zeta)^{-1} = -\sum_{h=1}^{s'} [(\zeta - \lambda'_h)^{-1} P'_h + (\zeta - \lambda'_h)^{-2} D'_h + \cdots + (\zeta - \lambda'_h)^{-N} D'^{N-1}_h];$$

this is easily verified by multiplying (5.39) from the left or the right by the expression for $T - \zeta$ given above (note that $D'^N_h = D'^N = 0$ since D' is nilpotent). Since the decomposition of an operator-valued meromorphic function into the sum of partial fractions is unique (just as for numerical functions), comparison of (5.39) with (5.23) shows that s', λ'_h, P'_h and D'_h must coincide with s, λ_h, P_h and D_h respectively. This completes the proof of the uniqueness of the spectral representation.

The spectral representation (5.36) leads to the *Jordan canonical form* of T. For this it is only necessary to recall the structure of the nilpotents D_h (see § 3.4). If we introduce a suitable basis in each M_h, the part D_{h, M_h} of D_h in M_h is represented by a matrix of the form (3.30). Thus T_{M_h} is represented by a triangular matrix $(\tau^{(h)}_{j\,k})$ of the form (3.30) with all the diagonal elements replaced by λ_h. If we collect all the basis elements of $\mathsf{M}_1, \ldots, \mathsf{M}_s$ to form a basis of X, the matrix of T is the direct sum of submatrices $(\tau^{(h)}_{j\,k})$. In particular, it follows that (see Problem 3.14)

$$(5.40) \qquad \det(T - \zeta) = \prod_{h=1}^{s} (\lambda_h - \zeta)^{m_h}, \quad \operatorname{tr} T = \sum_{h=1}^{s} m_h \lambda_h .$$

It is often convenient to count each eigenvalue λ_h of T repeatedly m_h times (m_h is the *algebraic* multiplicity of λ_h) and thus to denote the eigenvalues by $\mu_1, \mu_2, \ldots, \mu_N$ (for example $\mu_1 = \cdots = \mu_{m_1} = \lambda_1$, $\mu_{m_1+1} = \cdots = \mu_{m_1+m_2} = \lambda_2, \ldots$). For convenience we shall call $\mu_1, \ldots, \mu_N$ the *repeated eigenvalues of* T. Thus every operator $T \in \mathscr{B}(\mathsf{X})$ has exactly N repeated eigenvalues. (5.40) can be written

$$(5.41) \qquad \det(T - \zeta) = \prod_{k=1}^{N} (\mu_k - \zeta), \quad \operatorname{tr} T = \sum_{k=1}^{N} \mu_k .$$

Problem 5.12. The geometric eigenspace N_h of T for the eigenvalue λ_h is a subset of the algebraic eigenspace M_h. $\mathsf{N}_h = \mathsf{M}_h$ holds if and only if λ_h is semisimple.

Problem 5.13. λ_h is semisimple if and only if $\zeta = \lambda_h$ is a simple pole (pole of order 1) of $R(\zeta)$.

Problem 5.14. If n is sufficiently large ($n \geqq m$, say), rank T^n is equal to $N - m$, where m is the algebraic multiplicity of the eigenvalue 0 of T (it being agreed to set $m = 0$ if 0 is not an eigenvalue of T, see § 1.1).

Problem 5.15. We have

$$(5.42) \qquad |\operatorname{tr} T| \leqq (\operatorname{rank} T) \, \|T\| \leqq N \, \|T\| .$$

Problem 5.16. The eigenvalues λ_h of T are identical with the roots of the algebraic equation of degree N *(characterictic equation)*

$$(5.43) \qquad \det(T - \zeta) = 0 .$$

The multiplicity of λ_h as the root of this equation is equal to the algebraic multiplicity of the eigenvalue λ_h.

Let (τ_{jk}) be an $N \times N$ matrix. It can be regarded as the representation of a linear operator T in a vector space X, say the space C^N of N-dimensional numerical vectors. This means that the matrix representing T with respect to the canonical basis $\{x_j\}$ of X coincides with (τ_{jk}). Let (5.36) be the spectral representation of T, and let $\{x_j'\}$ be the basis of X used in the preceding paragraph to bring the associated matrix (τ_{jk}') of T into the canonical form.

The relationship between the bases $\{x_j\}$ and $\{x_j'\}$ is given by (1.2) and (1.4). Since the x_k' are eigenvectors of T in the generalized sense, the numerical vectors $(\hat{\gamma}_{1k}, \ldots, \hat{\gamma}_{Nk})$, $k = 1, \ldots, N$, are generalized eigenvectors of the matrix (τ_{jk}). The relationship between the matrices (τ_{jk}) and (τ_{jk}') is given by (3.8). Thus the transformation of (τ_{jk}) into the simpler form (τ_{jk}') is effected by the matrix $(\hat{\gamma}_{jk})$, constructed from the generalized eigenvectors of (τ_{jk}), according to the formula $(\tau_{jk}') = (\hat{\gamma}_{jk})^{-1} (\tau_{jk}) (\hat{\gamma}_{jk})$.

If the eigenvalue λ_h is semisimple, the h-th submatrix of (τ_{jk}') is a diagonal matrix with all diagonal elements equal to λ_h. If T is diagonable, (τ_{jk}') is itself a diagonal matrix with the diagonal elements $\lambda_1, \ldots, \lambda_s$, where λ_h is repeated m_h times (that is, with the diagonal elements $\mu_1, \ldots, \mu_N$, where the μ_j are the repeated eigenvalues). In this case $(\hat{\gamma}_{1k}, \ldots, \hat{\gamma}_{Nk})$, $k = 1, \ldots, N$, are eigenvectors of (τ_{jk}) in the proper sense.

Problem 5.17. $(\gamma_{j1}, \ldots, \gamma_{jN})$, $j = 1, \ldots, N$, are generalized eigenvectors of the transposed matrix of (τ_{jk}).

5. The adjoint problem

If $T \in \mathscr{B}(\mathsf{X})$, then $T^* \in \mathscr{B}(\mathsf{X}^*)$. There is a simple relationship between the spectral representations of the two operators T, T^*. If (5.36) is the spectral representation of T, then that of T^* is given by

$$T^* = S^* + D^* = \sum_{h=1}^{s} (\bar{\lambda}_h P_h^* + D_h^*) , \tag{5.44}$$

for the P_h^* as well as the P_h satisfy the relations (3.27–3.28) and the D_h^* are nilpotents commuting with the P_h^* and with one another (note the uniqueness of the spectral representation, par. 4). In particular, it follows that *the eigenvalues of T^* are the complex conjugates of the eigenvalues of T, with the same algebraic multiplicities.* The corresponding geometric multiplicities are also equal; this is a consequence of (3.40).

Problem 5.18. $R^*(\zeta) = R(\zeta, T^*) = (T^* - \zeta)^{-1}$ has the following properties:

$$R(\zeta)^* = R^*(\bar{\zeta}) , \tag{5.45}$$

$$R^*(\zeta) = - \sum_{h=1}^{s} \left[(\zeta - \bar{\lambda}_h)^{-1} P_h^* + \sum_{n=1}^{m_h - 1} (\zeta - \bar{\lambda}_h)^{-n-1} D_h^{*n} \right] . \tag{5.46}$$

6. Functions of an operator

If $p(\zeta)$ is a polynomial in ζ, the operator $p(T)$ is defined for any $T \in \mathscr{B}(\mathsf{X})$ (see § 3.3). Making use of the resolvent $R(\zeta) = (T - \zeta)^{-1}$, we can now define functions $\phi(T)$ of T for a more general class of functions $\phi(\zeta)$. It should be noted that $R(\zeta_0)$ is itself equal to $\phi(T)$ for $\phi(\zeta) = (\zeta - \zeta_0)^{-1}$.

Suppose that $\phi(\zeta)$ is holomorphic in a domain Δ of the complex plane containing all the eigenvalues λ_h of T, and let $\Gamma \subset \Delta$ be a simple closed smooth curve with positive direction enclosing all the λ_h in its interior. Then $\phi(T)$ is defined by the *Dunford-Taylor integral*

$$\phi(T) = -\frac{1}{2\pi i}\int_\Gamma \phi(\zeta)\, R(\zeta)\, d\zeta = \frac{1}{2\pi i}\int_\Gamma \phi(\zeta)\, (\zeta - T)^{-1}\, d\zeta\,. \tag{5.47}$$

This is an analogue of the Cauchy integral formula[1] in function theory. More generally, Γ may consist of several simple closed curves Γ_k with interiors Δ'_k such that the union of the Δ'_k contains all the eigenvalues of T. Note that (5.47) does not depend on Γ as long as Γ satisfies these conditions.

It is easily verified that (5.47) coincides with (3.18) when $\phi(\zeta)$ is a polynomial. It suffices to verify this for monomials $\phi(\zeta) = \zeta^n$, $n = 0, 1, 2, \ldots$. The proof for the case $n = 0$ is contained in (5.25). For $n \geqq 1$, write $\zeta^n = (\zeta - T + T)^n = (\zeta - T)^n + \cdots + T^n$ and substitute it into (5.47); all terms except the last vanish on integration by Cauchy's theorem, while the last term gives T^n.

Problem 5.19. If $\phi(\zeta) = \sum \alpha_n \zeta^n$ is an entire function, then $\phi(T) = \sum \alpha_n T^n$.

The correspondence $\phi(\zeta) \to \phi(T)$ is a homomorphism of the algebra of holomorphic functions on Δ into the algebra $\mathscr{B}(\mathsf{X})$:

$$\phi(\zeta) = \alpha_1 \phi_1(\zeta) + \alpha_2 \phi_2(\zeta) \text{ implies } \phi(T) = \alpha_1 \phi_1(T) + \alpha_2 \phi_2(T)\,, \tag{5.48}$$

$$\phi(\zeta) = \phi_1(\zeta)\, \phi_2(\zeta) \text{ implies } \phi(T) = \phi_1(T)\, \phi_2(T)\,. \tag{5.49}$$

This justifies the notation $\phi(T)$ for the operator defined by (5.47). The proof of (5.48) is obvious. For (5.49), it suffices to note that the proof is exactly the same as the proof of (5.17) for $n, m < 0$ (so that $\eta_n + \eta_m - 1 = -1$). In fact (5.17) is a special case of (5.49) for $\phi_1(\zeta) = \zeta^{-n-1}$ and $\phi_2(\zeta) = \zeta^{-m-1}$.

The spectral representation of $\phi(T)$ is obtained by substitution of (5.23) into (5.47). If $T = \sum (\lambda_h P_h + D_h)$ is the spectral representation of T, the result is

$$\phi(T) = \sum_{h=1}^{s} [\phi(\lambda_h)\, P_h + D'_h] \tag{5.50}$$

where[2]

[1] See KNOPP [[1]], p. 61.

[2] Note that $(2\pi i)^{-1} \int_\Gamma (\zeta - \lambda)^{-n-1} \phi(\zeta)\, d\zeta = \phi^{(n)}(\lambda)/n!$ if λ is inside Γ.

$$D'_h = \phi'(\lambda_h)\, D_h + \cdots + \frac{\phi^{(m_h-1)}(\lambda_h)}{(m_h-1)!}\, D_h^{m_h-1}\,. \tag{5.51}$$

Since the D'_h are nilpotents commuting with each other and with the P_k, it follows from the uniqueness of the spectral representation that (5.50) is the spectral representation of $\phi(T)$.

Thus $\phi(T)$ has the same eigenprojections P_h as T, and the eigenvalues of $\phi(T)$ are $\phi(\lambda_h)$ with multiplicities m_h. Strictly speaking, these statements are true only if the $\phi(\lambda_h)$ are different from one another. If $\phi(\lambda_1) = \phi(\lambda_2)$, for example, we must say that $\phi(\lambda_1)$ is an eigenvalue of $\phi(T)$ with the eigenprojection $P_1 + P_2$ and multiplicity $m_1 + m_2$.

That the $\phi(\lambda_h)$ are exactly the eigenvalues of $\phi(T)$ is a special case of the so-called *spectral mapping theorem.*

(5.48) and (5.49) are to be supplemented by another functional relation

$$\phi(\zeta) = \phi_1(\phi_2(\zeta)) \quad \text{implies} \quad \phi(T) = \phi_1(\phi_2(T))\,. \tag{5.52}$$

Here it is assumed that ϕ_2 is holomorphic in a domain Δ_2 of the sort stated before, whereas ϕ_1 is holomorphic in a domain Δ_1 containing all the eigenvalues $\phi_2(\lambda_h)$ of $S = \phi_2(T)$. This ensures that $\phi_1(S)$ can be constructed by

$$\phi_1(S) = \frac{1}{2\pi i} \int_{\Gamma_1} \phi_1(z)\,(z - S)^{-1}\, dz \tag{5.53}$$

where Γ_1 is a curve (or a union of curves) enclosing the eigenvalues $\phi_2(\lambda_h)$ of S. But we have

$$(z - S)^{-1} = (z - \phi_2(T))^{-1} = \frac{1}{2\pi i} \int_{\Gamma_2} (z - \phi_2(\zeta))^{-1} (\zeta - T)^{-1}\, d\zeta \tag{5.54}$$

for $z \in \Gamma_1$; this follows from (5.49) and the fact that both $z - \phi_2(\zeta)$ and $(z - \phi_2(\zeta))^{-1}$ are holomorphic in ζ in an appropriate subdomain of Δ_2 containing all the λ_h (the curve Γ_2 should be taken in such a subdomain so that the image of Γ_2 under ϕ_2 lies inside Γ_1). Substitution of (5.54) into (5.53) gives

$$\begin{aligned} \phi_1(S) &= \left(\frac{1}{2\pi i}\right)^2 \int_{\Gamma_1}\int_{\Gamma_2} \phi_1(z)\,(z - \phi_2(\zeta))^{-1} (\zeta - T)^{-1}\, d\zeta\, dz \\ &= \frac{1}{2\pi i} \int_{\Gamma_2} \phi_1(\phi_2(\zeta))\,(\zeta - T)^{-1}\, d\zeta = \phi(T) \end{aligned} \tag{5.55}$$

as we wished to show.

Example 5.20. As an application of the Dunford integral, let us define the logarithm

$$S = \log T \tag{5.56}$$

of an operator $T \in \mathscr{B}(X)$, assuming that T is nonsingular. We can take a simply connected domain Δ in the complex plane containing all the eigenvalues λ_h of T

but not containing 0. Let Γ be a simple closed curve in Δ enclosing all the λ_h. Since $\phi(\zeta) = \log\zeta$ can be defined as a holomorphic function on Δ, the application of (5.47) defines a function

$$S = \log T = \phi(T) = -\frac{1}{2\pi i}\int_\Gamma \log\zeta\, R(\zeta)\, d\zeta\,. \tag{5.57}$$

Since $\exp(\log\zeta) = \zeta$, it follows from (5.52) that

$$\exp(\log T) = T\,. \tag{5.58}$$

It should be noted that the choice of the domain Δ and the function $\phi(\zeta) = \log\zeta$ is not unique. Hence there are different operators $\log T$ with the above porperties. In particular, each of the eigenvalues of $\log T$ with one choice may differ by an integral multiple of $2\pi i$ from those with a different choice. If m_h is the algebraic multiplicity of λ_h, it follows that

$$\operatorname{tr}(\log T) = \sum m_h \log\lambda_h + 2n\pi i\,, \quad n = \text{integer}\,, \tag{5.59}$$

$$\exp(\operatorname{tr}(\log T)) = \prod \lambda_h^{m_h} = \det T\,. \tag{5.60}$$

7. Similarity transformations

Let U be an operator from a vector space X to another one X' such that $U^{-1} \in \mathscr{B}(\mathsf{X}', \mathsf{X})$ exists (this implies that $\dim\mathsf{X} = \dim\mathsf{X}'$). For an operator $T \in \mathscr{B}(\mathsf{X})$, the operator $T' \in \mathscr{B}(\mathsf{X}')$ defined by

$$T' = U T U^{-1} \tag{5.61}$$

is said to result from T by *transformation* by U. T' is said to be *similar*[1] to T. T' has the same internal structure as T, for the one-to-one correspondence $u \to u' = Uu$ is invariant in the sense that $Tu \to T'u' = UTu$. If we choose bases $\{x_j\}$ and $\{x_j'\}$ of X and X', respectively, in such a way that $x_j' = Ux_j$, then T and T' are represented by the same matrix.

If

$$T = \sum (\lambda_h P_h + D_h) \tag{5.62}$$

is the spectral representation of T, then

$$T' = \sum (\lambda_h P_h' + D_h')\,, \quad P_h' = U P_h U^{-1}\,, \quad D_h' = U D_h U^{-1}\,, \tag{5.63}$$

is the spectral representation of T'.

§ 6. Operators in unitary spaces

1. Unitary spaces

So far we have been dealing with general operators without introducing any special assumptions. In applications, however, we are often concerned with Hermitian or normal operators. Such notions are defined

[1] We have considered in § 4.6 an example of pairs of mutually similar operators.

only in a special kind of normed space H, called a *unitary space*, in which is defined an *inner product* (u, v) for any two vectors u, v. In this section we shall see what these notions add to our general results, especially in eigenvalue problems. We assume that $0 < \dim \mathsf{H} < \infty$.

The inner product (u, v) is complex-valued, (Hermitian) symmetric:

$$\overline{(u, v)} = (v, u) \tag{6.1}$$

and *sesquilinear*, that is, linear in u and semilinear in v. (6.1) implies that (u, u) is always real; it is further assumed to be positive-definite:

$$(u, u) > 0 \quad \text{for} \quad u \neq 0\,^{1}. \tag{6.2}$$

We shall see in a moment that the inner product (u, v) may be regarded as a special case of the scalar product defined between an element of a vector space and an element of its adjoint space. This justifies the use of the same symbol (u, v) for the two quantities.

Since a unitary space H is a vector space, there could be defined different norms in H. There exists, however, a distinguished norm in H *(unitary norm)* defined in terms of the inner product, and it is this one which is meant whenever we speak of the norm in a unitary space. [Of course any other possible norms are equivalent to this particular norm, see (1.20).] The unitary norm is given by

$$\|u\| = (u, u)^{1/2}. \tag{6.3}$$

Obviously the first two of the conditions (1.18) for a norm are satisfied. Before verifying the third condition (triangle inequality), we note the *Schwarz inequality*

$$|(u, v)| \leqq \|u\|\, \|v\|, \tag{6.4}$$

where equality holds if and only if u and v are linearly dependent. (6.4) follows, for example, from the identity

$$\|\, \|v\|^2 u - (u, v)\, v\|^2 = (\|u\|^2 \|v\|^2 - |(u, v)|^2)\, \|v\|^2. \tag{6.5}$$

The triangle inequality is a consequence of (6.4): $\|u + v\|^2 = (u + v, u + v) = \|u\|^2 + 2\,\mathrm{Re}\,(u, v) + \|v\|^2 \leqq \|u\|^2 + 2\|u\|\,\|v\| + \|v\|^2 = (\|u\| + \|v\|)^2$.

Example 6.1. For numerical vectors $u = (\xi_1, \ldots, \xi_N)$ and $v = (\eta_1, \ldots, \eta_N)$ set

$$(u, v) = \sum \xi_j \bar{\eta}_j, \quad \|u\|^2 = \sum |\xi_j|^2. \tag{6.6}$$

With this inner product the space $\mathbf{C}^N$ of N-dimensional numerical vectors becomes a unitary space.

Problem 6.2. The unitary norm has the characteristic property

$$\|u + v\|^2 + \|u - v\|^2 = 2\|u\|^2 + 2\|v\|^2. \tag{6.7}$$

Problem 6.3. The inner product (u, v) can be expressed in terms of the norm by

$$(u, v) = \frac{1}{4}\,(\|u + v\|^2 - \|u - v\|^2 + i\|u + i\,v\|^2 - i\|u - i\,v\|^2). \tag{6.8}$$

---

[1] In any finite-dimensional vector space X, one can introduce a positive-definite sesquilinear form and make X into a unitary space.

Problem 6.4. For any pair (ξ_j), (η_j) of numerical vectors, we have

$$(6.9)\qquad \begin{gathered} |\Sigma\, \xi_j\, \eta_j| \leqq (\Sigma\, |\xi_j|^2)^{\frac{1}{2}}\, (\Sigma\, |\eta_j|^2)^{\frac{1}{2}}\,, \\ (\Sigma\, |\xi_j + \eta_j|^2)^{\frac{1}{2}} \leqq (\Sigma\, |\xi_j|^2)^{\frac{1}{2}} + (\Sigma\, |\eta_j|^2)^{\frac{1}{2}}\,. \end{gathered}$$

2. The adjoint space

A characteristic property of a unitary space H is that the adjoint space H^* can be identified with H itself: $\mathsf{H}^* = \mathsf{H}$.

For any $u \in \mathsf{H}$ set $f_u[v] = (u, v)$. f_u is a semilinear form on H so that $f_u \in \mathsf{H}^*$. The map $u \to f_u$ is linear, for if $u = \alpha_1 u_1 + \alpha_2 u_2$ then $f_u[v] = (u, v) = \alpha_1 (u_1, v) + \alpha_2 (u_2, v) = \alpha_1 f_{u_1}[v] + \alpha_2 f_{u_2}[v]$ so that $f_u = \alpha_1 f_{u_1} + \alpha_2 f_{u_2}$. Thus $u \to f_u = T u$ defines a linear operator T on H to H^*.

T is isometric: $\|T u\| = \|u\|$. In fact $\|T u\| = \|f_u\| = \sup_v |(u, v)|/\|v\| = \|u\|$ by (2.22) [note the Schwarz inequality and $(u, u) = \|u\|^2$]. In particular T is one to one. Since $\dim \mathsf{H}^* = \dim \mathsf{H}$, it follows that T maps H onto the whole of H^* isometrically. This means that every $f \in \mathsf{H}^*$ has the form f_u with a uniquely determined $u \in \mathsf{H}$ such that $\|f\| = \|u\|$. It is natural to identify f with this u. It is in this sense that we identify H^* with H.

Since $(u, v) = f[v] = (f, v)$, the inner product (u, v) is seen to coincide with the scalar product (f, v).

We can now take over various notions defined for the scalar product to the inner product (see § 2.2). If $(u, v) = 0$ we write $u \perp v$ and say that u, v are mutually *orthogonal* (or *perpendicular*). u is orthogonal to a subset S of H, in symbol $u \perp \mathsf{S}$, if $u \perp v$ for all $v \in \mathsf{S}$. The set of all $u \in \mathsf{H}$ such that $u \perp \mathsf{S}$ is the annihilator of S and is denoted by $\mathsf{S}^\perp$. Two subsets S, S' of H are *orthogonal*, in symbol $\mathsf{S} \perp \mathsf{S}'$, if every $u \in \mathsf{S}$ and every $v \in \mathsf{S}'$ are orthogonal.

Problem 6.5. (u, v) is a continuous function of u, v.

Problem 6.6. The Pythagorean theorem:

$$(6.10)\qquad \|u + v\|^2 = \|u\|^2 + \|v\|^2 \quad \text{if} \quad u \perp v\,.$$

Problem 6.7. $u \perp \mathsf{S}$ implies $u \perp \mathsf{M}$ where M is the span of S. $\mathsf{S}^\perp$ is a linear manifold and $\mathsf{S}^\perp = \mathsf{M}^\perp$.

Consider two unitary spaces H and H'. A complex-valued function $\mathfrak{t}[u, u']$ defined for $u \in \mathsf{H}$ and $u' \in \mathsf{H}'$ is called a *sesquilinear form* on $\mathsf{H} \times \mathsf{H}'$ if it is linear in u and semilinear in u'. If in particular $\mathsf{H}' = \mathsf{H}$, we speak of a sesquilinear form on H. The inner product of H is a special case of a sesquilinear form on H. For a general sesquilinear form $\mathfrak{t}[u, v]$ on H, there is no relation between $\mathfrak{t}[u, v]$ and $\mathfrak{t}[v, u]$ so that the *quadratic form* $\mathfrak{t}[u] \equiv \mathfrak{t}[u, u]$ need not be real-valued. In any case, however, we have the relation *(polarization principle)*

$$(6.11)\qquad \mathfrak{t}[u, v] = \frac{1}{4}\left(\mathfrak{t}[u + v] - \mathfrak{t}[u - v] + i\, \mathfrak{t}[u + i\, v] - i\, \mathfrak{t}[u - i\, v]\right)$$

similar to (6.8). Thus the sesquilinear form $\mathfrak{t}[u, v]$ is determined by the associated quadratic form $\mathfrak{t}[u]$. In particular $\mathfrak{t}[u, v] = 0$ identically if $\mathfrak{t}[u] = 0$ identically. $\mathfrak{t}[u, v]$ is called the *polar form* of $\mathfrak{t}[u]$.

Problem 6.8. If $|\mathfrak{t}[u]| \leqq M\|u\|^2$ for all $u \in \mathsf{H}$, then $|\mathfrak{t}[u, v]| \leqq 2M\|u\| \|v\|$.

Problem 6.9. If T is a linear operator on H to H', $\|Tu\|^2$ is a quadratic form on H with the polar form (Tu, Tv).

Remark 6.10. The validity of (6.11) is closely related to the existence of the scalar i. The quadratic form $\mathfrak{t}[u]$ does *not* determine $\mathfrak{t}[u, v]$ in a *real* vector space.

3. Orthonormal families

A family of vectors $x_1, \ldots, x_n \in \mathsf{H}$ is called an *orthogonal family* if any two elements of this family are orthogonal. It is said to be *orthonormal* if, in addition, each vector is normalized:

$$(x_j, x_k) = \delta_{jk} \,. \tag{6.12}$$

As is easily seen, the vectors of an orthonormal family are linearly independent. An orthonormal family $\{x_1, \ldots, x_n\}$ is *complete* if $n = N = \dim \mathsf{H}$. Thus it is a basis of H, called an *orthonormal basis*.

Let M be the span of an orthonormal family $\{x_1, \ldots, x_n\}$. For any $u \in \mathsf{H}$, the vector

$$u' = \sum_{j=1}^{n} (u, x_j)\, x_j \tag{6.13}$$

has the property that $u' \in \mathsf{M}$, $u - u' \in \mathsf{M}^\perp$. u' is called the *orthogonal projection* (or simply the projection) of u on M. The Pythagorean theorem (6.10) gives

$$\|u\|^2 = \|u'\|^2 + \|u - u'\|^2 = \sum_{j=1}^{n} |(u, x_j)|^2 + \|u - u'\|^2 \,, \tag{6.14}$$

and hence

$$\sum_{j=1}^{n} |(u, x_j)|^2 \leqq \|u\|^2 \quad (\textit{Bessel's inequality}) \,. \tag{6.15}$$

For any linearly independent vectors $u_1, \ldots, u_n$, it is possible to construct an orthonormal family $x_1, \ldots, x_n$ such that for each $k = 1, \ldots, n$, the k vectors $x_1, \ldots, x_k$ span the same linear manifold as the k vectors $u_1, \ldots, u_k$. This is proved by induction on k. Suppose that $x_1, \ldots, x_{k-1}$ have been constructed. If M_{k-1} denotes the span of $x_1, \ldots, x_{k-1}$, then by hypothesis M_{k-1} is identical with the span of $u_1, \ldots, u_{k-1}$. Set $u_k'' = u_k - u_k'$ where u_k' is the projection of u_k on M_{k-1} (if $k = 1$ set $u_1'' = u_1$). The linear independence of the u_j implies that $u_k'' \neq 0$. Set $x_k = \|u_k''\|^{-1} u_k''$. Then the vectors $x_1, \ldots, x_k$ satisfy the required conditions. The construction described here is called the *Schmidt orthogonalization process*.

Since every linear manifold M of H has a basis $\{u_j\}$, it follows that M has an orthonormal basis $\{x_j\}$. In particular there exists a complete orthonormal family in H. An arbitrary vector x_1 with $\|x_1\| = 1$ can be the first element of an orthonormal basis of H.

It follows also that any $u \in \mathsf{H}$ has an orthogonal projection u' on a given linear manifold M. u' is determined uniquely by the property that $u' \in \mathsf{M}$ and $u'' = u - u' \in \mathsf{M}^\perp$. Thus H is the direct sum of M and $\mathsf{M}^\perp$:

$$\mathsf{H} = \mathsf{M} \oplus \mathsf{M}^\perp . \tag{6.16}$$

In this sense $\mathsf{M}^\perp$ is called the *orthogonal complement* of M. We have

$$\mathsf{M}^{\perp\perp} = \mathsf{M}\,, \quad \dim \mathsf{M}^\perp = N - \dim \mathsf{M} . \tag{6.17}$$

When $\mathsf{N} \subset \mathsf{M}$, $\mathsf{M} \cap \mathsf{N}^\perp$ is also denoted by $\mathsf{M} \ominus \mathsf{N}$; it is the set of all $u \in \mathsf{M}$ such that $u \perp \mathsf{N}$.

In the particular case $\mathsf{M} = \mathsf{H}$, we have $\mathsf{M}^\perp = 0$ and so $u'' = 0$. Thus (6.13) gives

$$u = \sum_{j=1}^{N} (u, x_j)\, x_j . \tag{6.18}$$

This is the *expansion* of u in the orthonormal basis $\{x_j\}$. Multiplication of (6.18) from the right by v gives

$$(u, v) = \sum_j (u, x_j)\, (x_j, v) = \sum_j (u, x_j)\, \overline{(v, x_j)} . \tag{6.19}$$

In particular

$$\|u\|^2 = \sum_j |(u, x_j)|^2 \quad (\textit{Parseval's equality}) . \tag{6.20}$$

The following lemma will be required later.

Lemma 6.11. *Let* M, M' *be two linear manifolds of* H *with dimensions* n, n' *respectively. Then* $\dim(\mathsf{M}' \cap \mathsf{M}^\perp) \geqq n' - n$.

This follows from (1.10) in view of the relations $\dim \mathsf{M}' = n'$, $\dim \mathsf{M}^\perp = N - n$, $\dim(\mathsf{M}' + \mathsf{M}^\perp) \leqq N$.

Let $\{x_j\}$ be a (not necessarily orthonormal) basis of H. The adjoint basis $\{e_j\}$ is a basis of $\mathsf{H}^* = \mathsf{H}$ satisfying the relations

$$(e_j, x_k) = \delta_{jk} \tag{6.21}$$

(see § 2.3). $\{x_j\}$ and $\{e_j\}$ are also said to form a *biorthogonal family* of elements of H. The basis $\{x_j\}$ is *selfadjoint* if $e_j = x_j$, $j = 1, \ldots, N$. Thus *a basis of* H *is selfadjoint if and only if it is orthonormal.*

4. Linear operators

Consider a linear operator T on a unitary space H to another unitary space H'. We recall that the norm of T is defined by $\|T\| = \sup \|T u\|/\|u\| = \sup |(T u, u')|/\|u\|\, \|u'\|$ [see (4.1), (4.2)].

The function

$$t[u, u'] = (Tu, u') \tag{6.22}$$

is a sesquilinear form on $H \times H'$. Conversely, an arbitrary sesquilinear form $t[u, u']$ on $H \times H'$ can be expressed in this form by a suitable choice of an operator T on H to H'. Since $t[u, u']$ is a semilinear form on H' for a fixed u, there exists a unique $w' \in H'$ such that $t[u, u'] = (w', u')$ for all $u' \in H'$. Since w' is determined by u, we can define a function T by setting $w' = Tu$. It can be easily seen that T is a linear operator on H to H'. If in particular t is a sesquilinear form on H $(H' = H)$, then T is a linear operator on H to itself.

In the same way, $t[u, u']$ can also be expressed in the form

$$t[u, u'] = (u, T^* u'), \tag{6.23}$$

where T^* is a linear operator on H' to H, called the *adjoint (operator)* of T. T^* coincides with the adjoint operator defined in § 3.6 by the identification of H^*, H'^* with H, H' respectively.

T^*T is a linear operator on H to itself. The relation

$$(u, T^*Tv) = (T^*Tu, v) = (Tu, Tv) \tag{6.24}$$

shows that T^*T is the operator associated with the sesquilinear form (Tu, Tv) on H. Note that the first two members of (6.24) are the inner product in H while the last is that in H'. It follows from (6.24) and (4.2) that $\|T^* T\| = \sup |(Tu, Tv)|/\|u\| \|v\| \geqq \sup \|Tu\|^2/\|u\|^2 = \|T\|^2$. Since, on the other hand, $\|T^*T\| \leqq \|T^*\| \|T\| = \|T\|^2$, we have

$$\|T^*T\| = \|T\|^2. \tag{6.25}$$

In particular, $T^*T = 0$ implies $T = 0$.

Problem 6.12. $N(T^* T) = N(T)$.

Problem 6.13. We have the polarization principle for an operator T on H to H

$$\begin{aligned}(Tu, v) = \frac{1}{4} [&(T(u + v), u + v) - (T(u - v), u - v) + \\ &+ i(T(u + iv), u + iv) - i(T(u - iv), u - iv)].\end{aligned} \tag{6.26}$$

Problem 6.14. If T is an operator on H to itself, $(Tu, u) = 0$ for all u implies $T = 0$.

The matrix representation of an operator is related to a pair of adjoint bases in the domain and range spaces [see (3.10)]. This suggests that the choice of selfadjoint (orthonormal) bases is convenient for the matrix representation of operators between unitary spaces.

Let T be an operator from a unitary space H to another one H' and let $\{x_k\}$, $\{x'_j\}$ be orthonormal bases of H, H', respectively. The matrix elements of T for these bases are then

$$\tau_{jk} = (T x_k, x'_j) = (x_k, T^* x'_j) \,, \tag{6.27}$$

as is seen from (3.10) by setting $f_j = x'_j$. More directly, (6.27) follows from the expansion

$$T x_k = \sum_j (T x_k, x'_j)\, x'_j \,. \tag{6.28}$$

Recall that if $\mathsf{H}' = \mathsf{H}$ it is the convention to take $x'_j = x_j$.

Problem 6.15. If $\{x_k\}$ is an orthonormal basis of H and if T is an operator on H to itself,

$$\operatorname{tr} T = \sum_k (T x_k, x_k) \,. \tag{6.29}$$

The matrix of T^* with respect to the same pair of bases $\{x'_j\}$, $\{x_k\}$ is given by $\tau^*_{kj} = (T^* x'_j, x_k)$. Comparison with (6.27) gives

$$\tau^*_{kj} = \overline{\tau_{jk}} \,. \tag{6.30}$$

Thus the matrices of T and T^* (for the same pair of orthogonal bases) are Hermitian conjugate to each other.

5. Symmetric forms and symmetric operators

A sesquilinear form $\mathfrak{t}[u, v]$ on a unitary space H is said to be *symmetric* if

$$\mathfrak{t}[v, u] = \overline{\mathfrak{t}[u, v]} \quad \text{for all} \quad u, v \in \mathsf{H} \,. \tag{6.31}$$

If $\mathfrak{t}[u, v]$ is symmetric, the associated quadratic form $\mathfrak{t}[u]$ is real-valued. The converse is also true, as is seen from (6.11).

A symmetric sesquilinear form (or the associated quadratic form) $\mathfrak{t}$ is *nonnegative* (in symbol $\mathfrak{t} \geqq 0$) if $\mathfrak{t}[u] \geqq 0$ for all u, and *positive* if $\mathfrak{t}[u] > 0$ for $u \neq 0$. The Schwarz and triangle inequalities are true for any nonnegative form as for the inner product (which is a special positive sesquilinear form):

$$\begin{aligned} |\mathfrak{t}[u, v]| &\leqq \mathfrak{t}[u]^{1/2}\, \mathfrak{t}[v]^{1/2} \leqq \frac{1}{2} (\mathfrak{t}[u] + \mathfrak{t}[v]) \,, \\ \mathfrak{t}[u + v]^{1/2} &\leqq \mathfrak{t}[u]^{1/2} + \mathfrak{t}[v]^{1/2} \,, \\ \mathfrak{t}[u + v] &\leqq 2\mathfrak{t}[u] + 2\mathfrak{t}[v] \,. \end{aligned} \tag{6.32}$$

Note that strict positivity was not used in the proof of similar inequalities for the inner product.

The *lower bound* γ of a symmetric form $\mathfrak{t}$ is defined as the largest real number such that $\mathfrak{t}[u] \geqq \gamma \|u\|^2$. The upper bound γ' is defined similarly. We have

$$|\mathfrak{t}[u, v]| \leqq M \|u\| \, \|v\| \,, \quad M = \max(|\gamma|, |\gamma'|) \,. \tag{6.33}$$

To see this, we note that the value $\mathfrak{t}[u, v]$ under consideration may be assumed to be real, for (6.33) is unchanged by multiplying u with a scalar of absolute value one. Since $\mathfrak{t}[u]$ is real-valued, we see from (6.11) that $\mathfrak{t}[u, v] = 4^{-1}(\mathfrak{t}[u+v] - \mathfrak{t}[u-v])$. Since $|\mathfrak{t}[u]| \leqq M\|u\|^2$ for all u, it follows that $|\mathfrak{t}[u, v]| \leqq 4^{-1}M(\|u+v\|^2 + \|u-v\|^2) = 2^{-1}M(\|u\|^2 + \|v\|^2)$. Replacement of u, v respectively by αu, v/α, with $\alpha^2 = \|v\|/\|u\|$, yields (6.33).

The operator T associated with a symmetric form $\mathfrak{t}[u, v]$ according to (6.22) has the property that

$$T^* = T \tag{6.34}$$

by (6.22), (6.23) and (6.31). An operator T on H to itself satisfying (6.34) is said to be *(Hermitian) symmetric* or *selfadjoint*. Conversely, a symmetric operator T determines a symmetric form $\mathfrak{t}[u, v] = (Tu, v)$ on H. Thus *(Tu, u) is real for all $u \in \mathsf{H}$ if and only if T is symmetric.* A symmetric operator T is *nonnegative (positive)* if the associated form is nonnegative (positive). For a nonnegative symmetric operator T we have the following inequalities corresponding to (6.32):

$$\begin{aligned} |(Tu, v)| &\leqq (Tu, u)^{1/2}(Tv, v)^{1/2}, \\ (T(u+v), u+v)^{1/2} &\leqq (Tu, u)^{1/2} + (Tv, v)^{1/2}. \end{aligned} \tag{6.35}$$

We write $T \geqq 0$ to denote that T is nonnegative symmetric. More generally, we write

$$T \geqq S \quad \text{or} \quad S \leqq T \tag{6.36}$$

if S, T are symmetric operators such that $T - S \geqq 0$. The upper and lower bounds of the quadratic form (Tu, u) are called the upper and lower bounds of the symmetric operator T.

Problem 6.16. If T is symmetric, $\alpha T + \beta$ is symmetric for real α, β. More generally, $p(T)$ is symmetric for any polynomial p with real coefficients.

Problem 6.17. For any linear operator T on H to H' (H, H' being unitary spaces), T^*T and TT^* are nonnegative symmetric operators in H and H', respectively.

Problem 6.18. If T is symmetric, then $T^2 \geqq 0$; $T^2 = 0$ if and only if $T = 0$. If T is symmetric and $T^n = 0$ for some positive integer n, then $T = 0$.

Problem 6.19. $R \leqq S$ and $S \leqq T$ imply $R \leqq T$. $S \leqq T$ and $S \geqq T$ imply $S = T$.

6. Unitary, isometric and normal operators

Let H and H' be unitary spaces. An operator T on H to H' is said to be *isometric*[1] if

$$\|Tu\| = \|u\| \quad \text{for every} \quad u \in \mathsf{H}. \tag{6.37}$$

[1] Isometric operators can be defined more generally between any two normed spaces, but we shall have no occasion to consider general isometric operators.

This is equivalent to $((T^*T - 1)\,u, u) = 0$ and therefore (see Problem 6.14)

$$T^*T = 1\,. \tag{6.38}$$

This implies that

$$(Tu, Tv) = (u, v) \quad \text{for every} \quad u, v \in \mathsf{H}\,. \tag{6.39}$$

An isometric operator T is said to be *unitary* if the range of T is the whole space H'. Since (6.37) implies that the mapping by T is one to one, it is necessary for the existence of a unitary operator on H to H' that $\dim \mathsf{H}' = \dim \mathsf{H}$. Conversely, if $\dim \mathsf{H}' = \dim \mathsf{H} < \infty$ any isometric operator on H to H' is unitary. As we shall see later, this is not true for infinite-dimensional spaces.

Problem 6.20. A $T \in \mathscr{B}(\mathsf{H}, \mathsf{H}')$ is unitary if and only if $T^{-1} \in \mathscr{B}(\mathsf{H}', \mathsf{H})$ exists and

$$T^{-1} = T^*\,. \tag{6.40}$$

Problem 6.21. T is unitary if and only if T^* is.

Problem 6.22. If $T \in \mathscr{B}(\mathsf{H}', \mathsf{H}'')$ and $S \in \mathscr{B}(\mathsf{H}, \mathsf{H}')$ are isometric, $TS \in \mathscr{B}(\mathsf{H}, \mathsf{H}'')$ is isometric. The same is true if "isometric" is replaced by "unitary".

Symmetric operators and unitary operators on a unitary space into itself are special cases of *normal operators*. $T \in \mathscr{B}(\mathsf{H})$ is said to be normal if T and T^* commute:

$$T^*T = T\,T^*\,. \tag{6.41}$$

This is equivalent to (again note Problem 6.14)

$$\|T^*u\| = \|Tu\| \quad \text{for all} \quad u \in \mathsf{H}\,. \tag{6.42}$$

An important property of a normal operator T is that

$$\|T^n\| = \|T\|^n\,, \quad n = 1, 2, \ldots\,. \tag{6.43}$$

This implies in particular that (spr denotes the spectral radius, see § 4.2)

$$\operatorname{spr} T = \|T\|\,. \tag{6.44}$$

To prove (6.43), we begin with the special case in which T is symmetric. We have then $\|T^2\| = \|T\|^2$ by (6.25). Since T^2 is symmetric, we have similarly $\|T^4\| = \|T^2\|^2 = \|T\|^4$. Proceeding in the same way, we see that (6.43) holds for $n = 2^m$, $m = 1, 2, \ldots$. Suppose now that T is normal but not necessarily symmetric. Again by (6.25) we have $\|T^n\|^2 = \|T^{n*}T^n\|$. But since $T^{n*}T^n = (T^*T)^n$ by (6.41) and T^*T is symmetric, we have $\|T^n\|^2 = \|(T^*T)^n\| = \|T^*T\|^n = \|T\|^{2n}$ for $n = 2^m$. This proves (6.43) for $n = 2^m$. For general n, we take an m such that $2^m - n = r \geqq 0$. Since (6.43) has been proved for n replaced by $n + r = 2^m$, we have $\|T\|^{n+r} = \|T^{n+r}\| \leqq \|T^n\|\,\|T^r\| \leqq \|T^n\|\,\|T\|^r$ or $\|T\|^n \leqq \|T^n\|$. But since the opposite inequality is obvious, (6.43) follows.

Problem 6.23. If T is normal, $p(T)$ is normal for any polynomial p.

Problem 6.24. T^{-1} is normal if T is normal and nonsingular.

Problem 6.25. If T is normal, $T^n = 0$ for some integer n implies $T = 0$. In other words, a normal operator T is nilpotent if and only if $T = 0$.

7. Projections

An important example of a symmetric operator is an *orthogonal projection*. Consider a subspace M of H and the decomposition $\mathsf{H} = \mathsf{M} \oplus \mathsf{M}^{\perp}$ [see (6.16)]. The projection operator $P = P_{\mathsf{M}}$ on M along $\mathsf{M}^{\perp}$ is called the orthogonal projection on M. P is *symmetric and nonnegative*, for (with the notation of par. 3)

$$(6.45) \qquad (Pu, u) = (u', u' + u'') = (u', u') \geqq 0$$

in virtue of $u' \perp u''$. Thus

$$(6.46) \qquad P^* = P , \quad P \geqq 0 , \quad P^2 = P .$$

Conversely, it is easy to see that a symmetric, idempotent operator $P \in \mathscr{B}(\mathsf{H})$ is an orthogonal projection on $\mathsf{M} = \mathsf{R}(P)$.

Problem 6.26. $1 - P$ is an orthogonal projection with P. If P is an orthogonal projection, we have

$$(6.47) \qquad 0 \leqq P \leqq 1^{1} , \quad \|P\| = 1 \quad \text{if} \quad P \neq 0 .$$

Problem 6.27. $\|(1 - P_{\mathsf{M}}) u\| = \operatorname{dist}(u, \mathsf{M}) , \quad u \in \mathsf{H}$.

Problem 6.28. $\mathsf{M} \perp \mathsf{N}$ is equivalent to $P_{\mathsf{M}} P_{\mathsf{N}} = 0$. The following three conditions are equivalent: $\mathsf{M} \supset \mathsf{N}$, $P_{\mathsf{M}} \geqq P_{\mathsf{N}}$, $P_{\mathsf{M}} P_{\mathsf{N}} = P_{\mathsf{N}}$.

Let $P_1, \ldots, P_n$ be orthogonal projections such that

$$(6.48) \qquad P_j P_k = \delta_{jk} P_j .$$

Then their sum

$$(6.49) \qquad P = \sum_{j=1}^{n} P_j$$

is also an orthogonal projection, of which the range is the direct sum of the ranges of the P_j.

Orthogonal projections are special kinds of projections. We can of course consider more general "oblique" projections in a unitary space H. Let $\mathsf{H} = \mathsf{M} \oplus \mathsf{N}$ with M, N not necessarily orthogonal, and let P be the projection on M along N. Then P^* is the projection on $\mathsf{N}^{\perp}$ along $\mathsf{M}^{\perp}$ [see (3.43)].

Problem 6.29. $\|P\| \geqq 1$ for a projection $P \neq 0$; $\|P\| = 1$ holds if and only if P is an orthogonal projection. [hint for "only if" part: Let $u \in \mathsf{N}^{\perp}$. Then $u \perp (1 - P) u$. Apply the Pythagorean theorem to obtain $\|Pu\|^2 = \|u\|^2 + \|(1 - P) u\|^2$. From

[1] The notation $0 \leqq P \leqq 1$, which is used to denote the order relation defined for symmetric operators [see (6.36)], does not conflict with the notation introduced earlier for projections (see § 3.4).

this and $\|P\| = 1$ deduce $\mathsf{N}^\perp \subset \mathsf{M}$. Consider P^* to deduce the opposite inclusion.]

Problem 6.30. A normal projection is orthogonal.

Problem 6.31. If P is a projection in a unitary space with $0 \neq P \neq 1$, then $\|P\| = \|1 - P\|$[1].

8. Pairs of projections

Let us now consider a pair P, Q of projections in H and recall the results of § 4.6 that $\mathsf{R}(P)$ and $\mathsf{R}(Q)$ are isomorphic if $P - Q$ is sufficiently small. A new result here is that *the operator U given by* (4.38) *is unitary if P, Q are orthogonal projections.* This is seen by noting that $U'^* = V'$ and $R^* = R$ [see (4.36) and (4.33)], which imply $U^* = V = U^{-1}$. Thus

Theorem 6.32. *Two orthogonal projections P, Q such that $\|P - Q\| < 1$ are unitarily equivalent, that is, there is a unitary operator U with the property $Q = U P U^{-1}$.*

Problem 6.33. $\|P - Q\| \leq 1$ for any pair P, Q of orthogonal projections [see (4.34)].

A similar but somewhat deeper result is given by the following theorem.

Theorem 6.34.[2] *Let P, Q be two orthogonal projections with $\mathsf{M} = \mathsf{R}(P)$, $\mathsf{N} = \mathsf{R}(Q)$ such that*

$$\|(1 - Q) P\| = \delta < 1 . \tag{6.50}$$

Then there are the following alternatives. Either

i) *Q maps M onto N one to one and bicontinuously, and*

$$\|P - Q\| = \|(1 - P) Q\| = \|(1 - Q) P\| = \delta ; \quad \textit{or} \tag{6.51}$$

ii) *Q maps M onto a proper subspace N_0 of N one to one and bicontinuously and, if Q_0 is the orthogonal projection on N_0,*

$$\begin{aligned} \|P - Q_0\| = \|(1 - P) Q_0\| = \|(1 - Q_0) P\| = \|(1 - Q) P\| = \delta , \\ \|P - Q\| = \|(1 - P) Q\| = 1 . \end{aligned} \tag{6.52}$$

Proof. For any $u \in \mathsf{M}$, we have $\|u - Qu\| = \|(1 - Q) Pu\| \leq \delta \|u\|$ so that $\|Qu\| \geq (1 - \delta) \|u\|$. Thus the mapping $u \to Qu$ of M into N is one to one and bicontinuous (continuous with its inverse mapping). Therefore, the image $Q\mathsf{M} = \mathsf{N}_0$ of this mapping is a closed subspace of N. Let Q_0 be the orthogonal projection on N_0.

For any $w \in \mathsf{H}$, $Q_0 w$ is in N_0 and hence $Q_0 w = Qu$ for some $u \in \mathsf{M}$. If $Q_0 w \neq 0$ then $u \neq 0$ and, since $(1 - P) u = 0$,

---

[1] See Del Pasqua [1], T. Kato [13].

[2] See T. Kato [12], Lemma 221. This theorem is true even for dim $\mathsf{H} = \infty$. A similar but slightly weaker result was given earlier by Akhiezer and Glazman [1], § 34.

$$\|(1-P)\,Q_0 w\| = \|(1-P)\,Q u\| = \|(1-P)\,(Q u - \|Q u\|^2\,\|u\|^{-2}\,u)\| \\ \leqq \|Q\,u - \|Q\,u\|^2\,\|u\|^{-2}\,u\|. \tag{6.53}$$

Hence

$$\|(1-P)\,Q_0 w\|^2 \leqq \|Q u\|^2 - \|Q u\|^4\,\|u\|^{-2} = \\ = \|Q u\|^2\,\|u\|^{-2}(\|u\|^2 - \|Q u\|^2) = \|u\|^{-2}\,\|Q_0 w\|^2\,\|(1-Q)\,u\|^2 \leqq \\ \leqq \|u\|^{-2}\,\|w\|^2\,\|(1-Q)\,P u\|^2 \leqq \delta^2\,\|w\|^2\,. \tag{6.54}$$

This inequality is true even when $Q_0 w = 0$. Hence

$$\|(1-P)\,Q_0\| \leqq \delta = \|(1-Q)\,P\|\,. \tag{6.55}$$

For any $w \in \mathsf{H}$, we have now

$$\|(P-Q_0)\,w\|^2 = \|(1-Q_0)\,P w - Q_0(1-P)\,w\|^2 \\ = \|(1-Q_0)\,P w\|^2 + \|Q_0(1-P)\,w\|^2 \tag{6.56}$$

since the ranges of $1-Q_0$ and Q_0 are orthogonal. Noting that $P = P^2$ and $1 - P = (1-P)^2$, we see from (6.56) that

$$\|(P-Q_0)\,w\|^2 \leqq \|(1-Q_0)\,P\|^2\,\|P w\|^2 + \|Q_0(1-P)\|^2\,\|(1-P)\,w\|^2. \tag{6.57}$$

Since $Q_0 P = Q_0 Q P = Q P$ by the definition of Q_0, we have $\|(1-Q_0)\,P\| = \|(1-Q)\,P\| = \delta$, and $\|Q_0(1-P)\| = \|(Q_0(1-P))^*\| = \|(1-P)\,Q_0\| \leqq \leqq \delta$ by (6.55). Hence

$$\|(P-Q_0)\,w\|^2 \leqq \delta^2(\|P w\|^2 + \|(1-P)\,w\|^2) = \delta^2\,\|w\|^2\,. \tag{6.58}$$

This gives $\|P - Q_0\| \leqq \delta$. Actually we have equality here, for

$$\delta = \|(1-Q)\,P\| = \|(1-Q_0)\,P\| = \|(P-Q_0)\,P\| \leqq \|P-Q_0\| \leqq \delta\,. \tag{6.59}$$

The fact that $\|P - Q_0\| = \delta < 1$ implies that P maps $\mathsf{N}_0 = \mathsf{R}(Q_0)$ *onto* M one to one (see Problem 4.11). Applying the above result (6.55) to the pair P, Q replaced by Q_0, P, we thus obtain $\|(1-Q_0)\,P\| \leqq \|(1-P)Q_0\|$. Comparison with (6.59) then shows that we have equality also in (6.55).

If $\mathsf{N}_0 = \mathsf{N}$, this completes the proof of i). If $\mathsf{N}_0 \neq \mathsf{N}$, it remains to prove the last equality of (6.52). Let v be an element of N not belonging to N_0. Since $P\mathsf{N}_0 = \mathsf{M}$ as noted above, there is a $v_0 \in \mathsf{N}_0$ such that $P v_0 = P v$. Thus $w = v - v_0 \in \mathsf{N}$, $w \neq 0$ and $P w = 0$, so that $(P-Q)\,w = -w$ and $Q(1-P)\,w = Q w = w$. Thus $\|P - Q\| \geqq 1$ and $\|(1-P)\,Q\| = \|Q(1-P)\| \geqq 1$. Since we have the opposite inequalities (see Problem 6.33), this gives the desired result.

As an application of Theorem 6.34, we deduce an inequality concerning pairs of oblique and orthogonal projections.

Theorem 6.35. *Let P', Q' be two oblique projections in* H *and let* $\mathsf{M} = \mathsf{R}(P')$, $\mathsf{N} = \mathsf{R}(Q')$. *Let P, Q be the orthogonal projections on* M, N *respectively. Then*

$$\|P - Q\| \leqq \|P' - Q'\|\,. \tag{6.60}$$

Proof. Since $\|P - Q\|$ never exceeds 1 (see Problem 6.33), it suffices to consider the case $\|P' - Q'\| = \delta' < 1$. For any $u \in \mathsf{H}$, we have (see Problem 6.27) $\|(1 - Q) Pu\| = \operatorname{dist}(Pu, \mathsf{N}) \leqq \|Pu - Q' Pu\| = \|(P' - Q') Pu\| \leqq \delta' \|Pu\| \leqq \delta' \|u\|$. Hence $\|(1 - Q) P\| \leqq \delta' < 1$. Similarly we have $\|(1 - P) Q\| \leqq \delta' < 1$. Thus Theorem 6.34 is applicable, where the case ii) is excluded, so that $\|P - Q\| = \|(1 - P) Q\| \leqq \delta' = \|P' - Q'\|$ by (6.51).

Problem 6.36. If P', Q' are oblique projections with $\|P' - Q'\| < 1$, then there is a unitary operator U such that $U\mathsf{M} = \mathsf{N}$, $U^{-1}\mathsf{N} = \mathsf{M}$, where $\mathsf{M} = \mathsf{R}(P')$, $\mathsf{N} = \mathsf{R}(Q')$. (This proposition can be extended directly to the infinite-dimensional case, in which it is not trivial.)

Problem 6.37. Let $P'(\varkappa)$ be an oblique projection depending continuously on $\varkappa$ for $0 \leqq \varkappa \leqq 1$, and let $P(\varkappa)$ be the orthogonal projection on $\mathsf{M}(\varkappa) = \mathsf{R}(P'(\varkappa))$. Then $P(\varkappa)$ is also continuous in $\varkappa$, and there is a family of unitary operators $U(\varkappa)$, depending continuously on $\varkappa$, such that $U(\varkappa)\, \mathsf{M}(0) = \mathsf{M}(\varkappa)$, $U(\varkappa)^{-1}\, \mathsf{M}(\varkappa) = \mathsf{M}(0)$.

Remark 6.37a. Suppose P, Q are orthogonal projections. We recall that $R = (P - Q)^2$ commutes with P and Q (see § 4.6). This leads to a simultaneous spectral decomposition of P and Q. Since R is selfadjoint with $0 \leqq R \leqq 1$ (cf. Problem 6.33), all eigenvalues λ of R satisfy $0 \leqq \lambda \leqq 1$ Each eigenspace $\mathsf{M}(\lambda)$ reduces both P and Q. In $\mathsf{M}(\lambda)$, P and Q behave essentially like two one-dimensional projections in a two-dimensional space[1], their ranges having the angle θ if $\lambda = \sin^2 \theta$.

9. The eigenvalue problem

We now consider the eigenvalue problem for an operator in a unitary space H. For a general operator T, there is not much simplification to be gained by the fact that the underlying space is unitary; this is clear if one notes that any vector space may be made into a unitary space by introducing an inner product, whereas the eigenvalue problem can be formulated without reference to any inner product, even to any norm. The advantage of considering a unitary space appears when the operator T has some special property peculiar to the context of a unitary space, such as being symmetric, unitary or normal.

Theorem 6.38. *A normal operator is diagonable, and its eigenprojections are orthogonal projections.*

Proof. Let T be normal. Since T and T^* commute, we have

$$(6.61)\qquad (T - \zeta)(T^* - \zeta') = (T^* - \zeta')(T - \zeta) .$$

If ζ is not an eigenvalue of T and ζ' is not an eigenvalue of T^*, the inverse of (6.61) exists and we have

$$(6.62)\qquad R^*(\zeta')\, R(\zeta) = R(\zeta)\, R^*(\zeta'),$$

[1] For details see DAVIS [2].

where $R(\zeta)$ and $R^*(\zeta)$ are the resolvents of T and T^* respectively. (6.62) shows that these resolvents commute. In view of the expression (5.22) for the eigenprojection P_h associated with the eigenvalue λ_h of T, a double integration of (6.62) along appropriate paths in the ζ and ζ' planes yields the relation

$$P_k^* P_h = P_h P_k^*, \quad h, k = 1, \ldots, s; \tag{6.63}$$

recall that T^* has the eigenvalues $\bar{\lambda}_h$ and the associated eigenprojections P_h^* (§ 5.5). In particular P_h and P_h^* commute, which means that the P_h are normal. This implies that the P_h are orthogonal projections (see Problem 6.30):

$$P_h^* = P_h, \quad h = 1, \ldots, s. \tag{6.64}$$

The eigennilpotents are given by $D_h = (T - \lambda_h) P_h$ and $D_h^* = (T^* - \bar{\lambda}_h) P_h^* = (T^* - \bar{\lambda}_h) P_h$ by (5.26). Since P_h and T commute, $P_h = P_h^*$ commutes with T^* and T^* commutes with T, it follows that D_h commutes with D_h^*, that is, D_h is normal. As a normal nilpotent D_h must be zero (Problem 6.25). Thus T is diagonable.

The spectral representation of a normal operator T thus takes the form

$$\begin{gathered} T = \sum_{h=1}^{s} \lambda_h P_h, \quad T^* = \sum_{h=1}^{s} \bar{\lambda}_h P_h, \\ P_h^* = P_h, \quad P_h P_k = \delta_{hk} P_h, \quad \sum_{h=1}^{s} P_h = 1. \end{gathered} \tag{6.65}$$

T and T^* have the same set of eigenspaces, which are at the same time algebraic and geometric eigenspaces and which are orthogonal to one another. It follows further from (6.65) that

$$T^* T = T T^* = \sum_{h=1}^{s} |\lambda_h|^2 P_h. \tag{6.66}$$

Hence

$$\begin{aligned} \|Tu\|^2 = (T^* T u, u) &= \sum_h |\lambda_h|^2 (P_h u, u) \leq (\max |\lambda_h|^2) \sum (P_h u, u) = \\ &= (\max |\lambda_h|^2) \|u\|^2, \end{aligned}$$

which shows the $\|T\| \leq \max |\lambda_h|$. On the other hand $\|Tu\| = |\lambda_h| \|u\|$ for $u \in \mathsf{M}_h = \mathsf{R}(P_h)$. Thus we obtain

$$\|T\| = \max_h |\lambda_h| \tag{6.67}$$

for a normal operator T.

If we choose an orthonormal basis in each M_h, the elements of these bases taken together constitute an orthonormal basis of H. In other words, there is an orthonormal basis $\{\varphi_n\}$ of H such that

$$T \varphi_n = \mu_n \varphi_n, \quad n = 1, \ldots, N, \tag{6.68}$$

in which μ_n, $n = 1, \ldots, N$, are the repeated eigenvalues of T. The matrix of T for this basis has the elements

$$\tau_{jk} = (T\,\varphi_k, \varphi_j) = \mu_k\,\delta_{jk}; \tag{6.69}$$

this is a diagonal matrix with the diagonal elements μ_j.

Problem 6.39. An operator with the spectral representation (6.65) is normal.

Problem 6.40. A symmetric operator has only real eigenvalues. A normal operator with only real eigenvalues is symmetric.

Problem 6.41. Each eigenvalue of a unitary operator has absolute value one. A normal operator with this property is unitary.

Problem 6.42. A symmetric operator is nonnegative (positive) if and only if its eigenvalues are all nonnegative (positive). The upper (lower) bound of a symmetric operator is the largest (smallest) of its eigenvalues.

Problem 6.43. If T is normal, then

$$\|R(\zeta)\| = 1/\min_k |\zeta - \lambda_k| = 1/\mathrm{dist}(\zeta, \Sigma(T)), \tag{6.70}$$
$$\|S_h\| = 1/\min_{k \neq h} |\lambda_h - \lambda_k|,$$

where $R(\zeta)$ is the resolvent of T and S_h is as in (5.18).

10. The minimax principle

Let T be a symmetric operator in H. T is diagonable and has only real eigenvalues (Problem 6.40). Let

$$\mu_1 \leqq \mu_2 \leqq \cdots \leqq \mu_N \tag{6.71}$$

be the repeated eigenvalues of T arranged in the ascending order. For each subspace M of H set

$$\mu[\mathsf{M}] = \mu[T, \mathsf{M}] = \min_{\substack{u \in M \\ \|u\| = 1}} (Tu, u) = \min_{0 \neq u \in M} \frac{(Tu, u)}{\|u\|^2}. \tag{6.72}$$

The *minimax* (or rather *maximin*) *principle* asserts that

$$\mu_n = \max_{\mathrm{codim}\,\mathsf{M} = n-1} \mu[\mathsf{M}] = \max_{\mathrm{codim}\,\mathsf{M} \leqq n-1} \mu[\mathsf{M}], \tag{6.73}$$

where the max is to be taken over all subspaces M with the indicated property. (6.73) is equivalent to the following two propositions:

$$\mu_n \geqq \mu[\mathsf{M}] \quad \text{for any} \quad \mathsf{M} \quad \text{with} \quad \mathrm{codim}\,\mathsf{M} \leqq n-1; \tag{6.74}$$

$$\mu_n \leqq \mu[\mathsf{M}_0] \quad \text{for some} \quad \mathsf{M}_0 \quad \text{with} \quad \mathrm{codim}\,\mathsf{M}_0 = n-1. \tag{6.75}$$

Let us prove these separately.

Let $\{\varphi_n\}$ be an orthonormal basis with the property (6.68). Each $u \in \mathsf{H}$ has the expansion

$$u = \sum \xi_n\,\varphi_n, \quad \xi_n = (u, \varphi_n), \quad \|u\|^2 = \sum |\xi_n|^2, \tag{6.76}$$

in this basis. Then

(6.77) $$T u = \sum \xi_n T \varphi_n = \sum \mu_n \xi_n \varphi_n , \quad (T u, u) = \sum \mu_n |\xi_n|^2 .$$

Let M be any subspace with $\operatorname{codim} \mathsf{M} \leqq n - 1$. The n-dimensional subspace M' spanned by $\varphi_1, \ldots, \varphi_n$ contains a nonzero vector u in common with M (this is a consequence of Lemma 6.11, where M is to be replaced by $\mathsf{M}^\perp$). This u has the coefficients $\xi_{n+1}, \xi_{n+2}, \ldots$ equal to zero, so that $(T u, u) = \sum \mu_k |\xi_k|^2 \leqq \mu_n \sum |\xi_k|^2 = \mu_n \|u\|^2$. This proves (6.74).

Let M_0 be the subspace consisting of all vectors orthogonal to $\varphi_1, \ldots, \varphi_{n-1}$, so that $\operatorname{codim} \mathsf{M}_0 = n - 1$. Each $u \in \mathsf{M}_0$ has the coefficients $\xi_1, \ldots, \xi_{n-1}$ zero. Hence $(T u, u) \geqq \mu_n \sum |\xi_k|^2 = \mu_n \|u\|^2$, which implies (6.75).

The minimax principle is a convenient means for characterizing the eigenvalues μ_n without any reference to the eigenvectors. As an application of this principle, we shall prove the *monotonicity principles*.

Theorem 6.44. *If S, T are symmetric operators such that $S \leqq T$, then the eigenvalues of S are not larger than the corresponding eigenvalues of T, that is,*

(6.78) $$\mu_n [S] \leqq \mu_n [T] , \quad n = 1, \ldots, N .$$

Here $\mu_n [T]$ denotes the n-th eigenvalue of T in the ascending order as in (6.71).

The proof follows immediately from the minimax principle, for $S \leqq T$ implies $(S u, u) \leqq (T u, u)$ and therefore $\mu [S, \mathsf{M}] \leqq \mu (T, \mathsf{M})$ for any subspace M.

Problem 6.45. For every pair of symmetric operators S, T,

(6.79) $$\mu_1 [S] + \mu_1 [T] \leqq \mu_1 [S + T] \leqq \mu_1 [S] + \mu_N [T] .$$

Let M be a subspace of H with the orthogonal projection P. For any operator T in H, $S = P T P$ is called the *orthogonal projection of T on M*. M is invariant under S, so that we can speak of the eigenvalues of S in M (that is, of the part S_M of S in M). Note that S_M has $N - r$ repeated eigenvalues, where $r = \operatorname{codim} \mathsf{M}$.

Theorem 6.46. *Let T, S, S_M be as above. If T is symmetric, S and S_M are also symmetric, and*

(6.80) $$\mu_n [T] \leqq \mu_n [S_\mathsf{M}] \leqq \mu_{n+r} [T] , \quad n = 1, \ldots, N - r .$$

Proof. The symmetry of S and of S_M is obvious. $\mu_n [S_\mathsf{M}]$ is equal to $\max \mu [S_\mathsf{M}, \mathsf{M}']$ taken over all $\mathsf{M}' \subset \mathsf{M}$ such that the codimension of M' relative to M is equal to $n - 1$ ($\dim \mathsf{M}/\mathsf{M}' = n - 1$). But we have $\mu [S_\mathsf{M}, \mathsf{M}'] = \mu [S, \mathsf{M}'] = \mu [T, \mathsf{M}']$ because $(S_\mathsf{M} u, u) = (S u, u) = (T u, u)$ for any $u \in \mathsf{M}$, and $\dim \mathsf{M}/\mathsf{M}' = n - 1$ implies $\operatorname{codim} \mathsf{M}' = \dim \mathsf{H}/\mathsf{M}' = n + r - 1$. Therefore $\mu_n [S_\mathsf{M}]$ does not exceed $\mu_{n+r} [T] = \max \mu [T, \mathsf{M}']$

taken over all $M' \subset H$ with $\operatorname{codim} M' = n + r - 1$. This proves the second inequality of (6.80).

On the other hand we have $\mu_n[T] = \mu[T, M_0]$ where M_0 is the same as in (6.75). Hence $\mu_n[T] \leqq \mu[T, M_0 \cap M] = \mu[S_M, M_0 \cap M]$. But $M_0 \cap M$ has codimension not larger than $n - 1$ relative to M because $\operatorname{codim} M_0 = n - 1$. Thus $\mu[S_M, M_0 \cap M] \leqq \mu_n[S_M]$ by (6.74). This proves the first inequality of (6.80).

11. Dissipative operators and contraction semigroups

A linear operator A in H is said to be *dissipative* if

(6.81) $$\|(A - \xi)u\| \geqq \xi \|u\| \quad \text{for all} \quad u \in H \quad \text{and} \quad \xi > 0.$$

A is said to be *accretive* if $-A$ is dissipative[1].

If we introduce the resolvent $R(\zeta) = (A - \zeta)^{-1}$ of A, condition (6.81) is equivalent to

(6.82) $$\|R(\xi)\| \leqq 1/\xi \quad \text{for} \quad \xi > 0,$$

it being implied that all positive real numbers belong to the resolvent set $P(A)$.

(6.82) implies that $\|(1 - tA)^{-1}\| \leqq 1$ for $t > 0$. Thus it follows from Corollary 4.15 that

(6.83) $$\|e^{tA}\| \leqq 1 \quad \text{for} \quad t > 0.$$

The family e^{tA}, when restricted to $t \geqq 0$, forms a *semigroup* according to $e^{(t+s)A} = e^{tA} e^{sA}$ (see (4.21b)). (6.83) shows that e^{tA} forms a *contraction semigroup* if A is dissipative.

The converse is also true. To see this, we note the formula

(6.84) $$-R(\zeta) = \int_0^\infty e^{-\zeta t} e^{tA}\, dt \quad \text{for} \quad \operatorname{Re} \zeta > 0,$$

which says that $R(\zeta)$ is the *Laplace transform* of e^{tA}. (6.84) follows easily from $(d/dt)(e^{-\zeta t} e^{tA}) = (A - \zeta) e^{-\zeta t} e^{tA}$ on integration.

Theorem 6.47. *A is dissipative if and only if* $\operatorname{Re}(Au, u) \leqq 0$ *for all* $u \in H$.

Proof. (6.81) is equivalent to $\|Au\|^2 \geqq 2\xi \operatorname{Re}(Au, u)$ for all $\xi > 0$. Obviously this is the case if and only if $\operatorname{Re}(Au, u) \leqq 0$.

Theorem 6.48. *The following conditions are equivalent.*

(i) *A is dissipative.*
(ii) *$R(\xi)$ is dissipative for all $\xi > 0$.*
(iii) *$-AR(\xi)$ is dissipative for all $\xi > 0$.*

---

[1] Some of the following definitions and results (including (iii) of Theorem 6.49) are valid in any (∞-dimensional) Banach space X. More details on dissipative operators in a Hilbert space are given in V-§ 3.10–11.

Proof. First we note that if T^{-1} exists, T^{-1} is dissipative if and only if T is. This follows from Theorem 6.47, since $(T^{-1}v, v) = (u, Tu)$ for $u = T^{-1}v$. But A is dissipative if and only if $A - \xi$ is dissipative for all $\xi > 0$, as is readily seen from Theorem 6.47. Hence follows the equivalence of (i) and (ii). (iii) follows from (i) by $\operatorname{Re}(A R(\xi) v, v) = \operatorname{Re}(A u, (A - \xi) u) = \|A u\|^2 - \xi \operatorname{Re}(A u, u) \geqq 0$, where $u = R(\xi) v$. Conversely, (iii) implies (i) because $-\xi A R(\xi) \to A$ as $\xi \to \infty$.

Theorem 6.49. *Let A be dissipative. If 0 is an eigenvalue of A, it is semisimple, the associated eigenprojection P is selfadjoint, and the associated reduced resolvent S at 0 is dissipative.*

Proof. (6.82) shows that $R(\zeta)$ has at most a pole of order 1 at $\zeta = 0$. In view of the Laurent expansion formula (5.18), where we set $\lambda_h = 0$, $P_h = P$, $D_h = D$, we must have $D = 0$ (so that $\lambda = 0$ is semisimple) and $\|P\| \leqq 1$. It follows that $P^* = P$ (see Problem 6.29).

On the other hand, we have

$$S = \lim_{\xi \searrow 0} (1 - P) R(\xi) (1 - P)$$

by (5.18) and (5.19). Since $R(\xi)$ is dissipative for $\xi > 0$ by Theorem 6.48, (ii), we obtain

$$\operatorname{Re}(S u, u) = \lim_{\xi \searrow 0} \operatorname{Re}(R(\xi) v, v) \leqq 0, \quad v = (1 - P) u.$$

This proves that S is dissipative,

Problem 6.50. If H is symmetric, $\pm i H$ are both dissipative and accretive. If H is symmetric and $H \geqq 0$, then $c H$ is dissipative for $\operatorname{Re} c \leqq 0$; in particular H is accretive and $-H$ is dissipative.

§7. Positive matrices

1. Definitions and notation

An $N \times N$ matrix T is said to be positive [nonnegative] if all its matrix elements are positive [nonnegative] real numbers. This notion is entirely different from that of positive [nonnegative] symmetric operators considered in the preceding section (§ 6.5). In the present section we introduce basic facts concerning nonnegative matrices[1]. Some further

[1] For nonnegative matrices, see e.g. GANTMACHER [[1]], SCHAEFER [[1]]. To avoid possible confusions with the positivity (or positive-definiteness) of symmetric operators considered in § 6.5, we shall use the notations $T > 0$ and $T \geqq 0$ only in the sense of § 6.5. Sometimes nonnegative operators in the sense of the present section are said to be *positivity-preserving*. A convenient notation for it is $T \in \mathscr{B}_+(\mathsf{X})$. In practice there will be no possibility of confusion, however, since we shall be concerned only with the new kind of positivity throughout most of this section.

results related to perturbation of such matrices will be discussed in the next chapter.

In what follows we consider the complex vector space $\mathsf{X} = \mathbf{C}^N$, $0 < N < \infty$, the set of all complex *numerical* vectors $u = (\xi_j)$. u is said to be nonnegative [positive] if $\xi_j \geqq 0$ [$\xi_j > 0$] for $1 \leqq j \leqq N$, in symbol $u \geqq 0$ [$u \gg 0$]. The notation $u > 0$ will be reserved for "$u \geqq 0$ und $u \neq 0$". It should be noted that this definition is in terms of the canonical basis of X (see Example 1.4); it is not allowed to use components of u with respect to other bases of X.

It is convenient to introduce in X a norm $\| \ \|$, such as (1.15) to (1.17). We shall keep it unspecified unless necessary.

For $u = (\xi_j) \in \mathsf{X}$, we write

$$|u| = (|\xi_j|), \quad \text{so that} \quad |u| \geqq 0. \tag{7.1}$$

We define positivity also for vectors in the adjoint space X^*, which we identify with $\mathbf{C}^N$ in the usual way. Thus (see (2.19), (2.33))

$$(u, f) = \overline{(f, u)} = \sum_{j=1}^{N} \xi_j \bar{\alpha}_j, \quad u = (\xi_j) \in \mathsf{X}, \quad f = (\alpha_j) \in \mathsf{X}^*. \tag{7.2}$$

We say $f = (\alpha_j) \in \mathsf{X}^*$ is nonnegative [positive] if $\alpha_j \geqq 0$ [$\alpha_j > 0$] for all j, in symbol $f \geqq 0$ [$f \gg 0$]. Again $f > 0$ means $f \geqq 0$ and $f \neq 0$.

Lemma 7.1. *$f \geqq 0$ if and only if $(f, u) \geqq 0$ for all $u \geqq 0$. $f \gg 0$ if and only if $(f, u) > 0$ for all $u > 0$. $u \geqq 0$ if and only if $(f, u) \geqq 0$ for all $f \geqq 0$. $u \gg 0$ if and only if $(f, u) > 0$ for all $f > 0$.*

For any $f = (\alpha_j) \in \mathsf{X}^*$, we define $|f| = (|\alpha_j|)$ so that $|f| \geqq 0$. Obviously we have

$$|(u, f)| = |(f, u)| \leqq (|u|, |f|) = (|f|, |u|). \tag{7.3}$$

An $N \times N$ matrix T defines a linear operator in X in the usual way (see (3.6)), which we identify with T. We say T is *nonnegative* [*positive*] if all its matrix elements are nonnegative [positive]. An equivalent definition is that T is nonnegative if $u \geqq 0$ implies $Tu \geqq 0$, and T is positive if $u > 0$ implies $Tu \gg 0$. It is easy to see that T is nonnegative if and only if T^* is. If T is nonnegative, we have

$$|Tu| \leqq T|u| \ (u \in \mathsf{X}), \quad |T^* f| \leqq T^*|f| \quad (f \in \mathsf{X}^*), \tag{7.4}$$

$$|(Tu, f)| = |(u, T^* f)| \leqq (T|u|, |f|) = (|u|, T^*|f|). \tag{7.5}$$

The set of all nonnegative vectors in X [X^*] is denoted by X_+ [X^*_+]. X_+ [X^*_+] is a *total* (or *fundamental*) subset of X [X^*] in the sense that the linear combinations of vectors in X_+ [X^*_+] fill the whole space X [X^*_+].

2. The spectral properties of nonnegative matrices

The following results are part of the so-called *Perron-Frobenius theorem*.

Theorem 7.2. *Let T be a nonnegative matrix. Its spectral radius* spr T *is an eigenvalue of T with a nonnegative eigenvector u. No other eigenvalue of T exceeds* spr T *in absolute value or has a positive eigenvector.*

spr T is called the *principal* (or *dominant*) *eigenvalue* of T. It should be noted that other eigenvalues of T may also have nonnegative eigenvectors.

Proof. The resolvent $R(\zeta) = (T - \zeta)^{-1}$ has the expansion (5.10):

$$R(\zeta) = -\sum_{n=0}^{\infty} \zeta^{-n-1} T^n \quad (|\zeta| > r = \text{spr } T). \tag{7.6}$$

Since T is nonnegative, all its powers T^n are nonnegative. It follows that $-R(\xi)$ is nonnegative for real $\xi > r$. Since $(d/d\zeta)^k R(\zeta) = k!\ R(\zeta)^{k+1}$ by (5.8), we have

$$-(-d/d\xi)^k R(\xi) \text{ is nonnegative for } \xi > r. \tag{7.7}$$

For any $u \in \mathsf{X}$ and $f \in \mathsf{X}^*$, (7.6) gives

$$|(R(\zeta)\, u, f)| \leq \sum_{n=0}^{\infty} |\zeta|^{-n-1} |(T^n u, f)| \tag{7.8}$$

$$\leq \sum_{n=0}^{\infty} |\zeta|^{-n-1} (T^n |u|, |f|) = -(R(|\zeta|)\, |u|, |f|) \quad (|\zeta| > r),$$

where we have used (7.5).

According to (5.12), there is at least a pole $r e^{i\theta}$ of the resolvent $R(\zeta)$, where $0 \leq \theta < 2\pi$. Hence there are $u \in \mathsf{X}$ and $f \in \mathsf{X}^*$ such that $|(R(\zeta)\, u, f)| \to \infty$ as $\zeta \to r e^{i\theta}$. Then (7.8) shows that $-(R(|\zeta|)\, |u|, |f|) \to \infty$ as $|\zeta| \searrow r$. This implies that r is a pole of $R(\zeta)$, hence an eigenvalue of T.

Now we consider the Laurent expansion of $R(\zeta)$ at $\zeta = r$:

$$R(\zeta) = -(\zeta - r)^{-m} D^{m-1} - \cdots - (\zeta - r)^{-1} P + S + (\zeta - r) S^2 + \cdots, \tag{7.9}$$

(see (5.18)). Since $-R(\xi)$ is nonnegative for $\xi > r$ as noted above, (7.9) shows that $D^{m-1} = -\lim_{\xi \searrow r} (\xi - r)^m R(\xi)$ is nonnegative. Since $D^{m-1} \neq 0$ (due to the definition of m), one can find $v > 0$ such that $u = D^{m-1} v > 0$ (recall that X_+ is total). Since $(T - r)\, u = (T - r)\, D^{m-1}\, v = D^m\, v = 0$ (see (5.30) and note that $D^m = 0$), u is a nonnegative eigenvector of T for the eigenvalue r.

To prove the last statement of the theorem, let μ be another eigenvalue of T, with an eigenvector v. Since T^* is nonnegative with T, spr $T^* =$ spr $T = r$ is an eigenvalue of T^* with an eigenvector $u^* > 0$. Thus $\mu(v, u^*) = (Tv, u^*) = (v, T^* u^*) = r(v, u^*)$. Since $\mu \neq r$, it follows that $(v, u^*) = 0$, which is impossible if $v \gg 0$. [$|\mu| \leq r$ is known, see (5.12)].

Theorem 7.3. *If T is positive, the principal eigenvalue of T is positive and simple (has algebraic multiplicity one). There are no other eigenvalues of T with a nonnegative eigenvector.*

Proof. Let $u > 0$ be an eigenvector of T for the eigenvalue $r = \operatorname{spr} T$. Since $ru = Tu \gg 0$ by $u > 0$, we must have $r > 0$ and $u \gg 0$.

Next we prove that r has geometric multiplicity one. Suppose v is another real eigenvector for r. Then we can find a real number c such that $v + cu$ is nonnegative but not positive. If $v + cu \neq 0$, it is a nonnegative eigenvector of T for r, hence must be positive by the result given above. Since this is a contradiction, we must have $v + cu = 0$ or $v = -cu$. Thus there is no eigenvector for r linearly independent of u.

Now we prove that $D = 0$ or, equivalently, $m = 1$ in (7.9). Suppose $m \geqq 2$. Then $(T - r)D^{m-2} = D^{m-1}$ by (5.30) (set $D^0 = P$). Hence $(T - r)w = u$ for $w = D^{m-2}v$ in the notation of the proof of Theorem 7.2; here we may assume $u \gg 0$ because the geometric multiplicity of r has been shown to be one. It follows that $(u, u^*) = ((T - r)w, u^*) = (w, (T^* - r)u^*) = 0$, a contradiction to $u \gg 0$ and $u^* > 0$.

$D = 0$ implies that the algebraic eigenspace $P\mathsf{X}$ coincides with the geometric eigenspace. Hence $\dim P = 1$ and r is simple.

The last statement of the theorem can also be proved by using u^*. Suppose $Tz = \mu z$ with $z > 0$. Then we obtain $\mu(z, u^*) = (Tz, u^*) = (z, T^*u^*) = r(z, u^*)$. Since $(z, u^*) > 0$ by $u^* \gg 0$, we must have $\mu = r$.

3. Semigroups of nonnegative operators

We recall that for any $T \in \mathscr{B}(\mathsf{X})$, e^{tT} forms a group of operators when t varies over all real or complex numbers, and a semigroup when t is restricted to nonnegative real numbers (see (4.21b) and § 6.11). It follows from the Taylor series (4.20) that if T is a nonnegative matrix, e^{tT} is also nonnegative for $t \geqq 0$. But the converse is not true.

To characterize the generator of a nonnegative semigroup, we introduce the following definition. A matrix T is *essentially nonnegative* [*positive*] if all the off-diagonal elements of T are nonnegative [positive].

Theorem 7.4. *e^{tT} is nonnegative for $t > 0$ if and only if T is essentially nonnegative.*

Proof. If e^{tT} is nonnegative, $e^{tT} - 1$ has nonnegative off-diagonal elements, since the unit matrix 1 has off-diagonal elements zero. Thus (4.21c) shows that T has off-diagonal elements nonnegative. If, conversely, T has nonnegative off-diagonal elements, $T + c$ is nonnegative if the real number c is sufficiently large. Then $e^{t(T+c)} = e^{ct}e^{tT}$ is nonnegative for $t > 0$. Hence e^{tT} is nonnegative too.

Theorem 7.5. *Let T be essentially nonnegative. T has a distinguished eigenvalue $\lambda[T]$ (hereafter called the* principal *or* dominant eigenvalue[1])

[1] $\lambda[T]$ coincides with the *spectral bound* of T, which is defined for any operator T as $\sup \operatorname{Re} \lambda$ for $\lambda \in \Sigma(T)$.

with the following properties. (a) $\lambda[T]$ *is real.* (b) $\lambda[T]$ *has a nonnegative eigenvector, while no other eigenvalues of* T *have a positive eigenvector.* (c) $\lambda[T]$ *is strictly larger than the real parts of the other eigenvalues of* T. (d) $\operatorname{spr} e^{tT} = e^{t\lambda[T]}$ *for* $t > 0$. (e) *If, in particular,* T *is nonnegative, then* $\lambda[T] = \operatorname{spr} T$.

Proof. Choose a real number c such that $T + c$ is nonnegative. If we set $\lambda[T] = \operatorname{spr}(T + c) - c$, we see easily that $\lambda[T]$ is independent of c and satisfies the stated conditions (see Theorems 7.2 and 7.4). Note that e^{tT} has eigenvalues $e^{t\lambda_h}$ if T has eigenvalues λ_h, $h = 1, 2, \ldots, s$.

Problem 7.6. If T is essentially nonnegative and dissipative (X being assumed to be a unitary space), then $\lambda[T] \leqq 0$.

4. Irreducible matrices

First we prove a partial refinement of Theorem 7.4.

Theorem 7.7. *Let* T *be essentially nonnegative. Then the matrix elements of* e^{tT} *are entire functions of* t. *The diagonal elements of* e^{tT} *never vanish for* $t \geqq 0$, *while each off-diagonal element is either identically zero or strictly positive for* $t > 0$.

Proof. The elements of e^{tT} are entire functions of t for any $T \in \mathscr{B}(\mathsf{X})$. In view of the relation $e^{t(T+c)} = e^{ct} e^{tT}$, it suffices to prove the theorem when T is nonnegative. The Taylor expansion (4.20) then shows that the matrix elements of e^{tT} are convergent power series in t with nonnegative coefficients, so that they are nondecreasing real-analytic functions of $t \geqq 0$. Since $e^{0T} = 1$, the assertions of the theorem follow immediately.

An essentially nonnegative matrix T is said to be *irreducible* if e^{tT} is positive for all $t > 0$ or, equivalently (by Theorem 7.7), for some $t > 0$. Otherwise T is *reducible.*

Obviously, the reducibility of T does not depend on its diagonal part. T is irreducible if and only if T^* is. If T is irreducible and S is not smaller than T (i.e. $S - T$ is nonnegative), then S is irreducible. Thus the reducibility of T depends only on which off-diagonal elements of T are positive or zero, not on their size.

Problem 7.8. A nonnegative matrix T is said to be *primitive* if T^m is positive for some integer $m > 0$. If T is primitive, it is irreducible.

Remark 7.9. The origin of the term irreducible is the following. If a nonnegative matrix T is reducible, some of the matrix elements of e^{tT} are identically zero. Thus there is an index pair (j_0, k_0) such that the matrix elements $T^n(j_0, k_0) = 0$ for all $n = 0, 1, 2, \ldots$. Divide the index set $\{1, 2, \ldots, N\}$ into two disjoint subsets J and K in such a way that $T^n(j_0, k) = 0$ for all n if $k \in \mathrm{K}$, while this is not the case for $k \in \mathrm{J}$. Then both J and K are nonempty; indeed $j_0 \in \mathrm{J}$ and $k_0 \in \mathrm{K}$. Now we claim that $T(j, k) = 0$ whenever $j \in \mathrm{J}$ and $k \in \mathrm{K}$. To see this, choose n so that $T^n(j_0, j) \neq 0$. Then $0 = T^{n+1}(j_0, k) = \sum_{h=1}^{N} T^n(j_0, h)\, T(h, k)$ so that we must have

$T(j, k) = 0$. Thus the subspace of X spanned by the basis vectors e_k for $k \in \mathrm{K}$ is invariant under T, and T is reducible in that sense.

If the numbering of the e_j is rearranged so that $\mathrm{K} = \{1, 2, \ldots, s\}$ and $\mathrm{J} = \{s+1, \ldots, N\}$, the matrix T has all elements zero in the lower left rectangle of size $(n - s) \times s$. In other words, T is reducible if and only if it can be brought into such a form by a *simultaneous* rearrangement of rows and columns.

Theorem 7.10. *Let T be essentially nonnegative and irreducible. Then the principal eigenvalue $\lambda[T]$ (see Theorem 7.5) is simple, with a positive eigenvector. There are no other eigenvalues with a nonnegative eigenvector.*

Proof. Let λ_h, $h = 1, 2, \ldots, s$, be the eigenvalues of T, with the associated eigenprojections P_h. Then e^{tT} has exactly the eigenvalues $e^{t\lambda_h}$ with the same eigenprojections P_h (except when some of the $e^{t\lambda_h}$ coincide, which may be avoided if $t > 0$ is chosen sufficiently small). Since e^{tT} is positive for $t > 0$, spr (e^{tT}) is a simple eigenvalue of e^{tT} with a positive eigenvector u (Theorem 7.3). This implies that spr $(e^{tT}) = e^{t\lambda[T]}$ and $\lambda[T]$ is a simple eigenvalue of T with eigenvector $u \gg 0$ (see Theorem 7.5). The last statement of the theorem follows from the fact that no other eigenvalue of e^{tT} than spr (e^{tT}) has a nonnegative eigenvector (Theorem 7.3).

Example 7.11. (a) $T = \begin{pmatrix} 0 & 1 & 0 \\ 0 & 0 & 1 \\ 1 & 0 & 0 \end{pmatrix}$ is irreducible, though it is not primitive. $\lambda[T] = 1$, with an eigenvector (1,1,1).

b) $T = \begin{pmatrix} 1 & 1 & 1 \\ 0 & 1 & 1 \\ 0 & 1 & 1 \end{pmatrix}$ is reducible. $\lambda[T] = 2$ has an eigenvector (2,1,1). Another eigenvalue $\lambda_2 = 1$ has a nonnegative eigenvector (1,0,0). A third eigenvalue $\lambda_1 = 0$ has no nonnegative eigenvector.

Problem 7.12. If T is essentially nonnegative and irreducible, the same is true of T^*.

5. Positivity and dissipativity

As was remarked at the beginning of this section, positivity is a notion independent of positive-definiteness (or, more generally, dissipativity or accretivity). The former is related to concrete matrix representation of an operator, whereas the latter is related to the metric property of the underlying space (unitary structure). Nevertheless, there is a certain relationship between the two kinds of notions[1]. Here we use explicitly

[1] It should be noted that the two kinds of positivity may sometimes look contradictory to each other. For example, $T = \begin{pmatrix} -1 & 1 \\ 1 & -1 \end{pmatrix}$ is an essentially positive matrix (so that e^T is positive), but $T \leqq 0$ in the sense of § 6.5. If one goes to infinite-dimensional spaces X such as $\mathsf{L}^2(\mathrm{R}^n)$, the Laplacian $T = \Delta = \partial^2/\partial x_1^2 + \cdots + \partial^2/\partial x_n^2$ gives a typical example of the same phenomenon.

the unitary structure of X, introducing the unitary norm (1.17) into $X = C^N$ and identifying X^* with X.

Theorem 7.13. *Let T be essentially nonnegative and irreducible. Let $\lambda[T]$ be its principal eigenvalue. Then $T - \lambda[T]$ is similar to a dissipative operator with an eigenvalue zero. More precisely, there is a real diagonal matrix F with positive elements such that $F^{-1}TF - \lambda[T]$ is dissipative and has a simple eigenvalue zero, with a selfadjoint eigenprojection P.*

Proof. We may assume that T is nonnegative, by adding a suitable scalar to T if necessary. Then T^* is nonnegative irreducible too, and both T and T^* have positive eigenvectors for the principal eigenvalue $\lambda = \lambda[T] = \lambda[T^*] > 0$:

$$Tu = \lambda u, \quad T^* f = \lambda f, \quad u = (\xi_j) \gg 0, \quad f = (\alpha_j) \gg 0. \tag{7.10}$$

Set

$$w = ((\xi_j \alpha_j)^{1/2}) \gg 0, \quad F = \operatorname{diag}((\xi_j/\alpha_j)^{1/2}). \tag{7.11}$$

(F is a diagonal matrix with the diagonal elements indicated.) Then

$$u = Fw, \quad f = F^{-1}w, \tag{7.12}$$

so that (7.10) gives

$$F^{-1}TFw = \lambda w, \quad FT^*F^{-1}w = \lambda w. \tag{7.13}$$

If we set $B = F^{-1}TF$, (7.13) shows that B and B^* have the same eigenvalue λ with the same eigenvector w. Hence B^*B has eigenvalue λ^2 with the eigenvector $w \gg 0$. Since B^*B is nonnegative, λ^2 must be the principal eigenvalue of B^*B (see Theorem 7.5). Thus λ^2 is the largest eigenvalue of the *selfadjoint* operator B^*B, so that $\|B^*B\| = \lambda^2$ (see (6.67)). Since $\|B^*B\| = \|B\|^2$ by (6.25), it follows that

$$\|B\| = \lambda. \tag{7.14}$$

The dissipativity of $B - \lambda$ then follows from $\operatorname{Re}(Bx, x) \leqq \|Bx\| \|x\| \leqq \|B\| \|x\|^2 = \lambda \|x\|^2$ (see Theorem 6.47).

Since B is similar to T, the eigenvalue λ of B is simple by Theorem 7.10. The associated eigenprojection is selfadjoint by Theorem 6.49.

Problem 7.14. If T is essentially nonnegative and $T^* = T$, then $T - \lambda[T] \leqq 0$ in the sense of (6.36).

Chapter Two

Perturbation theory in a finite-dimensional space

In this chapter we consider perturbation theory for linear operators in a finite-dimensional space. The main question is how the eigenvalues and eigenvectors (or eigenprojections) change with the operator, in particular when the operator depends on a parameter analytically. This is a special case of a more general and more interesting problem in which the operator acts in an infinite-dimensional space.

The reason for discussing the finite-dimensional case separately is threefold. In the first place, it is not trivial. Second, it essentially embodies certain features of perturbation theory in the general case, especially those related to isolated eigenvalues. It is convenient to treat them in this simplified situation without being bothered by complications arising from the infinite dimensionality of the underlying space. The modifications required when going to the infinite-dimensional case will be introduced as supplements in later chapters, together with those features of perturbation theory which are peculiar to the infinite-dimensional case. Third, the finite-dimensional theory has its own interest, for example, in connection with the numerical analysis of matrices. The reader interested only in finite-dimensional problems can find what he wants in this chapter, without having to disentangle it from the general theory.

As mentioned above, the problem is by no means trivial, and many different methods of solving it have been introduced. The method used here is based on a function-theoretic study of the resolvent, in particular on the expression of eigenprojections as contour integrals of the resolvent. This is the quickest way to obtain general results as well as to deduce various estimates on the convergence rates of the perturbation series. In a certain sense the use of function theory for operator-valued functions is not altogether elementary, but since students of applied mathematics are as a rule well-acquainted with function theory, the author hopes that its presence in this form will not hinder those who might use the book for applications.

§ 1. Analytic perturbation of eigenvalues

1. The problem

We now go into one of our proper subjects, the perturbation theory for the eigenvalue problem in a finite-dimensional vector space X[1]. A typical problem of this theory is to investigate how the eigenvalues and eigenvectors (or eigenspaces) of a linear operator T change when T is

[1] In this chapter we assume that $0 < \dim \mathsf{X} = N < \infty$. Some of the results are valid in Banach spaces of infinite dimension, as will be noted each time when appropriate.

subjected to a small perturbation[1]. In dealing with such a problem, it is often convenient to consider a family of operators of the form

$$T(\varkappa) = T + \varkappa T' \tag{1.1}$$

where $\varkappa$ is a scalar parameter supposed to be small. $T(0) = T$ is called the *unperturbed operator* and $\varkappa T'$ the *perturbation*. A question arises whether the eigenvalues and the eigenvectors of $T(\varkappa)$ can be expressed as power series in $\varkappa$, that is, whether they are holomorphic functions of $\varkappa$ in the neighborhood of $\varkappa = 0$. If this is the case, the change of the eigenvalues and eigenvectors will be of the same order of magnitude as the perturbation $\varkappa T'$ itself for small $|\varkappa|$. As we shall see below, however, this is not always the case.

(1.1) can be generalized to

$$T(\varkappa) = T + \varkappa T^{(1)} + \varkappa^2 T^{(2)} + \cdots. \tag{1.2}$$

More generally, we may suppose that an operator-valued function $T(\varkappa)$ is given, which is holomorphic in a given domain D_0 of the complex $\varkappa$-plane[2].

The eigenvalues of $T(\varkappa)$ satisfy the characteristic equation (see Problem I-5.16)

$$\det(T(\varkappa) - \zeta) = 0. \tag{1.3}$$

This is an algebraic equation in ζ of degree $N = \dim \mathsf{X}$, with coefficients which are holomorphic in $\varkappa$; this is seen by writing (1.3) in terms of the matrix of $T(\varkappa)$ with respect to a basis $\{x_j\}$ of X, for each element of this matrix is a holomorphic function of $\varkappa$ [see I-(3.10)]. It follows from a well-known result in function theory that the roots of (1.3) are (branches of) *analytic functions of $\varkappa$ with only algebraic singularities*. More precisely, the roots of (1.3) for $\varkappa \in D_0$ constitute one or several branches of one or several analytic functions that have only algebraic singularities in D_0.

[1] For perturbation theory in general in a finite-dimensional space, see BAUMGÄRTEL ⟦1⟧, [4–7], DAVIS [1, 3], DAVIS and KAHAN [1], B. L. LIVŠIC [1], REED and SIMON ⟦2⟧, RELLICH [1, 8], VIŠIK and LYUSTERNIK [1, 3]. The book of BAUMGÄRTEL is a most comprehensive treatise, in particular for analytic perturbation theory. Numerically oriented readers may also consult PARLETT ⟦1⟧. Reference should be made also to basic papers and books dealing with analytic perturbation theory in infinite-dimensional spaces, such as COURANT and HILBERT ⟦1⟧, DUNFORD and SCHWARTZ ⟦1⟧, FRIEDRICHS ⟦1⟧, PORATH [1, 2], REED and SIMON ⟦2⟧, RELLICH [1–7], RIESZ and SZ.-NAGY ⟦1⟧, ROSENBLOOM [1], SCHÄFKE [3–5], SCHRÖDER [1–3], ŠMUL'YAN [1], T. KATO [1, 3, 6].

[2] One can restrict $\varkappa$ to real values, but since (1.2) given for real $\varkappa$ can always be extended to complex $\varkappa$, there is no loss of generality in considering complex $\varkappa$.

[3] See KNOPP ⟦2⟧, p. 119, where algebraic functions are considered, Actually (1.3) determines ζ as *algebroidal* (not necessarily algebraic) functions, which are, however, locally similar to algebraic functions. For detailed function-theoretic treatment of (1.3), see BAUMGÄRTEL ⟦1⟧, [1].

It follows immediately that the number of eigenvalues of $T(\varkappa)$ is a constant s independent of $\varkappa$, with the exception of some special values of $\varkappa$. There are only a finite number of such *exceptional points* $\varkappa$ in each compact subset of D_0. This number s is equal to N if these analytic functions (if there are more than one) are all distinct; in this case $T(\varkappa)$ is simple and therefore diagonable for all non-exceptional $\varkappa$. If, on the other hand, there happen to be identical ones among these analytic functions, then we have $s < N$; in this case $T(\varkappa)$ is said to be *permanently degenerate*.

Example 1.1. Here we collect the simplest examples illustrating the various possibilities stated above. These examples are concerned with a family $T(\varkappa)$ of the form (1.1) in a two-dimensional space ($N = 2$). For simplicity we identify $T(\varkappa)$ with its matrix representation with respect to a basis.

a) $$T(\varkappa) = \begin{pmatrix} 1 & \varkappa \\ \varkappa & -1 \end{pmatrix}.$$

The eigenvalues of $T(\varkappa)$ are

$$\lambda_{\pm}(\varkappa) = \pm (1 + \varkappa^2)^{1/2} \tag{1.4}$$

and are branches of one double-valued analytic function $(1 + \varkappa^2)^{1/2}$. Thus $s = N = 2$ and the exceptional points are $\varkappa = \pm i$, $T(\pm i)$ having only the eigenvalue 0.

b) $$T(\varkappa) = \begin{pmatrix} 0 & \varkappa \\ \varkappa & 0 \end{pmatrix}, \quad s = N = 2.$$

The eigenvalues are $\pm \varkappa$; these are two distinct entire functions of $\varkappa$ (the characteristic equation is $\zeta^2 - \varkappa^2 = 0$ and is reducible). There is one exceptional point $\varkappa = 0$, for which $T(\varkappa)$ has only one eigenvalue 0.

c) $$T(\varkappa) = \begin{pmatrix} 0 & \varkappa \\ 0 & 0 \end{pmatrix}, \quad s = 1.$$

$T(\varkappa)$ is permanently degenerate, the only eigenvalue being 0 for all $\varkappa$; we have two identical analytic functions zero. There are no exceptional points.

d) $$T(\varkappa) = \begin{pmatrix} 0 & 1 \\ \varkappa & 0 \end{pmatrix}, \quad s = 2.$$

The eigenvalues are $\pm \varkappa^{1/2}$, constituting one double-valued function $\varkappa^{1/2}$. There is one exceptional point $\varkappa = 0$.

e) $$T(\varkappa) = \begin{pmatrix} 1 & \varkappa \\ 0 & 0 \end{pmatrix}, \quad s = 2.$$

The eigenvalues are 0 and 1. There are no exceptional points.

f) $$T(\varkappa) = \begin{pmatrix} \varkappa & 1 \\ 0 & 0 \end{pmatrix}, \quad s = 2.$$

The eigenvalues are 0 and $\varkappa$, which are two distinct entire functions. There is one exceptional point $\varkappa = 0$.

2. Singularities of the eigenvalues

We now consider the eigenvalues of $T(\varkappa)$ in more detail. Since these are in general multiple-valued analytic functions of $\varkappa$, some care is needed in their notation. If $\varkappa$ is restricted to a simply-connected[1] subdomain D of the fundamental domain D_0 containing no exceptional

[1] See KNOPP [[1]], p. 19.

point (for brevity such a subdomain will be called a *simple subdomain*), the eigenvalues of $T(\varkappa)$ can be written

$$\lambda_1(\varkappa), \lambda_2(\varkappa), \ldots, \lambda_s(\varkappa), \tag{1.5}$$

all s functions $\lambda_h(\varkappa)$, $h = 1, \ldots, s$, being holomorphic in D and $\lambda_h(\varkappa) \neq \lambda_k(\varkappa)$ for $h \neq k$.

We next consider the behavior of the eigenvalues in the neighborhood of one of the exceptional points, which we may take as $\varkappa = 0$ without loss of generality. Let D be a small disk near $\varkappa = 0$ but excluding $\varkappa = 0$. The eigenvalues of $T(\varkappa)$ for $\varkappa \in$ D can be expressed by s holomorphic functions of the form (1.5). If D is moved continuously around $\varkappa = 0$, these s functions can be continued analytically. When D has been brought to its initial position after one revolution around $\varkappa = 0$, the s functions (1.5) will have undergone a permutation among themselves. These functions may therefore be grouped in the manner

$$\{\lambda_1(\varkappa), \ldots, \lambda_p(\varkappa)\}, \{\lambda_{p+1}(\varkappa), \ldots, \lambda_{p+q}(\varkappa)\}, \ldots, \tag{1.6}$$

in such a way that each group undergoes a cyclic permutation by a revolution of D of the kind described. For brevity each group will be called a *cycle* at the exceptional point $\varkappa = 0$, and the number of elements of a cycle will be called its *period*.

It is obvious that the elements of a cycle of period p constitute a branch of an analytic function (defined near $\varkappa = 0$) with a branch point (if $p \geqq 2$) at $\varkappa = 0$, and we have *Puiseux series* such as[1]

$$\lambda_h(\varkappa) = \lambda + \alpha_1 \omega^h \varkappa^{1/p} + \alpha_2 \omega^{2h} \varkappa^{2/p} + \cdots, \quad h = 0, 1, \ldots, p-1, \tag{1.7}$$

where $\omega = \exp(2\pi i/p)$. It should be noticed that here no negative powers of $\varkappa^{1/p}$ appear, for the coefficient of the highest power ζ^N in (1.3) is $(-1)^N$ so that the $\lambda_h(\varkappa)$ are continuous at $\varkappa = 0$[2]. $\lambda = \lambda_h(0)$ will be called the *center* of the cycle under consideration.

(1.7) shows that $|\lambda_h(\varkappa) - \lambda|$ is in general of the order $|\varkappa|^{1/p}$ for small $|\varkappa|$ for $h = 1, \ldots, p$. If $p \geqq 2$, therefore, the rate of change at an exceptional point of the eigenvalues of a cycle of period p is infinitely large compared with the change of $T(\varkappa)$ itself[3].

Problem 1.2. The sum of the $\lambda_h(\varkappa)$ belonging to a cycle is holomorphic at the exceptional point in question.

In general there are several cycles with the same center λ. All the eigenvalues (1.7) belonging to cycles with center λ are said to depart from the unperturbed eigenvalue λ by *splitting* at $\varkappa = 0$. The set of these

---

[1] See KNOPP [[2]], p. 130.
[2] See KNOPP [[2]], p. 122.
[3] This fact is of some importance in the numerical analysis of eigenvalues of matrices.

eigenvalues will be called the *λ-group*, since they cluster around λ for small $|\varkappa|$.

Remark 1.3. An exceptional point need not be a branch point of an analytic function representing some of the eigenvalues. In other words, it is possible that all cycles at an exceptional point $\varkappa = \varkappa_0$ are of period 1. In any case, however, some two different eigenvalues for $\varkappa \neq \varkappa_0$ must coincide at $\varkappa = \varkappa_0$ (definition of an exceptional point). Thus there is always splitting at (and only at) an exceptional point.

Example 1.4. Consider the examples listed in Example 1.1. We have a cycle of period 2 at the exceptional points $\varkappa = \pm i$ in a) and also at $\varkappa = 0$ in d). There are two cycles of period 1 at $\varkappa = 0$ in b) and f). There are no exceptional points in c) and e).

3. Perturbation of the resolvent

The resolvent

$$R(\zeta, \varkappa) = (T(\varkappa) - \zeta)^{-1} \tag{1.8}$$

of $T(\varkappa)$ is defined for all ζ not equal to any of the eigenvalues of $T(\varkappa)$ and is a meromorphic function of ζ for each fixed $\varkappa \in D_0$. Actually we have

Theorem 1.5. *$R(\zeta, \varkappa)$ is holomorphic in the two variables ζ, $\varkappa$ in each domain in which ζ is not equal to any of the eigenvalues of $T(\varkappa)$.*

Proof. Let $\zeta = \zeta_0$, $\varkappa = \varkappa_0$ belong to such a domain; we may assume $\varkappa_0 = 0$ without loss of generality. Thus ζ_0 is not equal to any eigenvalue of $T(0) = T$, and

$$\begin{aligned} T(\varkappa) - \zeta &= T - \zeta_0 - (\zeta - \zeta_0) + A(\varkappa) \\ &= [1 - (\zeta - \zeta_0 - A(\varkappa))\, R(\zeta_0)]\, (T - \zeta_0)\,, \end{aligned} \tag{1.9}$$

$$A(\varkappa) = T(\varkappa) - T = \sum_{n=1}^{\infty} \varkappa^n\, T^{(n)}\,, \tag{1.10}$$

where $R(\zeta) = R(\zeta, 0) = (T - \zeta)^{-1}$ and we assumed the Taylor expansion of $T(\varkappa)$ at $\varkappa = 0$ in the form (1.2). Hence

$$R(\zeta, \varkappa) = R(\zeta_0)\, [1 - (\zeta - \zeta_0 - A(\varkappa))\, R(\zeta_0)]^{-1}\,, \tag{1.11}$$

exists if the factor $[\;]^{-1}$ can be defined by a convergent Neumann series (see Example I-4.5), which is the case if, for example,

$$|\zeta - \zeta_0| + \sum_{n=1}^{\infty} |\varkappa|^n\, \|T^{(n)}\| < \|R(\zeta_0)\|^{-1}\,, \tag{1.12}$$

since $|\zeta - \zeta_0| + \|A(\varkappa)\|$ is not greater than the left member of (1.12). This inequality is certainly satisfied for sufficiently small $|\zeta - \zeta_0|$ and $|\varkappa|$, and then the right member of (1.11) can be written as a double power series in $\zeta - \zeta_0$ and $\varkappa$. This shows that $R(\zeta, \varkappa)$ is holomorphic in ζ and $\varkappa$ in a neighborhood of $\zeta = \zeta_0$, $\varkappa = 0$.

For later use it is more convenient to write $R(\zeta, \varkappa)$ as a power series in $\varkappa$ with coefficients depending on ζ. On setting $\zeta_0 = \zeta$ in (1.11), we obtain

$$(1.13)\quad R(\zeta,\varkappa) = R(\zeta)\,[1 + A(\varkappa)\,R(\zeta)]^{-1}$$
$$= R(\zeta)\sum_{p=0}^{\infty}[-A(\varkappa)\,R(\zeta)]^p = R(\zeta) + \sum_{n=1}^{\infty}\varkappa^n R^{(n)}(\zeta)\,,$$

where

$$(1.14)\quad R^{(n)}(\zeta) = \sum_{\substack{\nu_1+\cdots+\nu_p=n\\ \nu_j\geqq 1}} (-1)^p R(\zeta)\,T^{(\nu_1)}R(\zeta)\,T^{(\nu_2)}\ldots T^{(\nu_p)}R(\zeta)\,,$$

the sum being taken for all combinations of positive integers p and $\nu_1,\ldots,\nu_p$ such that $1\leqq p\leqq n$, $\nu_1+\cdots+\nu_p = n$.

(1.13) will be called the *second Neumann series* for the resolvent. It is uniformly convergent for sufficiently small $\varkappa$ and $\zeta\in\Gamma$ if Γ is a compact subset of the resolvent set $\mathrm{P}(T)$ of $T = T(0)$; this is seen from (1.12) with $\zeta_0 = \zeta$, where $\|R(\zeta)\|^{-1}$ has a positive minimum for $\zeta\in\Gamma$.

Theorem 1.5a. *Fix ζ in Theorem* 1.5. *Then either* i) $R(\zeta,\varkappa)$ *does not exist for any $\varkappa\in \mathrm{D}_0$, or* ii) $R(\zeta,\varkappa)$ *is meromorphic in $\varkappa\in\mathrm{D}_0$ (so that it exists for all $\varkappa$ except for isolated points).*

Proof. If $\det(T(\varkappa)-\zeta) = 0$ identically in $\varkappa$, we have the case i). Otherwise ii) follows easily from the matrix representation of $R(\zeta,\varkappa)$; note that the matrix elements are polynomials in those of $T(\varkappa)-\zeta$ divided by $\det(T(\varkappa)-\zeta)$, which is holomorphic in $\varkappa$.

Example 1.6. The resolvent for the $T(\varkappa)$ of Example 1.1, a) is given by

$$(1.15)\qquad R(\zeta,\varkappa) = (\zeta^2 - 1 - \varkappa^2)^{-1}\begin{pmatrix} -1-\zeta & -\varkappa\\ -\varkappa & 1-\zeta\end{pmatrix}.$$

Problem 1.7. Find the resolvents of the $T(\varkappa)$ of b) to f) in Example 1.1.

4. Perturbation of the eigenprojections and eigennilpotents

Let λ be one of the eigenvalues of $T = T(0)$, with multiplicity[1] m. Let Γ be a closed positively-oriented curve, say a circle, in the resolvent set $\mathrm{P}(T)$ enclosing λ but no other eigenvalues of T. As noted above, the second Neumann series (1.13) is then convergent for sufficiently small $|\varkappa|$ uniformly for $\zeta\in\Gamma$. The existence of the resolvent $R(\zeta,\varkappa)$ of $T(\varkappa)$ for $\zeta\in\Gamma$ implies that there are no eigenvalues of $T(\varkappa)$ on Γ.

The operator

$$(1.16)\qquad P(\varkappa) = -\frac{1}{2\pi i}\int_\Gamma R(\zeta,\varkappa)\,d\zeta \text{ [2]}$$

is a projection and is equal to the sum of the eigenprojections for all the eigenvalues of $T(\varkappa)$ lying inside Γ (see Problem I-5.9). In particular $P(0) = P$ coincides with the eigenprojection for the eigenvalue λ of T. Integrating (1.13) term by term, we have

[1] By "multiplicity" we mean the *algebraic* multiplicity unless otherwise stated.

[2] This integral formula is basic throughout the present book. In perturbation theory it was first used by Sz.-Nagy [1] and T. Kato [1], greatly simplifying the earlier method of Rellich [1]—[5].

(1.17) $$P(\varkappa) = P + \sum_{n=1}^{\infty} \varkappa^n P^{(n)}$$

with

(1.18) $$P^{(n)} = -\frac{1}{2\pi i} \int_\Gamma R^{(n)}(\zeta)\, d\zeta .$$

The series (1.17) is convergent for small $|\varkappa|$ so that $P(\varkappa)$ is holomorphic near $\varkappa = 0$. It follows from Lemma I-4.10 that the range $\mathsf{M}(\varkappa)$ of $P(\varkappa)$ is isomorphic with the (algebraic) eigenspace $\mathsf{M} = \mathsf{M}(0) = P\mathsf{X}$ of T for the eigenvalue λ. In particular we have

(1.19) $$\dim P(\varkappa) = \dim P = m .$$

Since (1.19) is true for all sufficiently small $|\varkappa|$, it follows that the eigenvalues of $T(\varkappa)$ lying inside Γ form exactly the λ-group. For brevity we call $P(\varkappa)$ the *total projection*, and $\mathsf{M}(\varkappa)$ the *total eigenspace*, for the λ-group.

If $\varkappa = 0$ is not an exceptional point, there is no splitting at $\varkappa = 0$ of the eigenvalue λ in question. In this case there is exactly one eigenvalue $\lambda(\varkappa)$ of $T(\varkappa)$ in the neighborhood of λ, and $P(\varkappa)$ is itself the eigenprojection for this eigenvalue $\lambda(\varkappa)$. (1.19) shows that the multiplicity of $\lambda(\varkappa)$ is equal to m. Similar results hold when $\varkappa = 0$ is replaced by any other non-exceptional point $\varkappa = \varkappa_0$.

Now consider a simple subdomain D of the $\varkappa$-plane and the set (1.5) of the eigenvalues of $T(\varkappa)$ for $\varkappa \in$ D, and let $P_h(\varkappa)$ be the eigenprojection for the eigenvalue $\lambda_h(\varkappa)$, $h = 1, \ldots, s$. The result just proved shows that each $P_h(\varkappa)$ is holomorphic in D and that each $\lambda_h(\varkappa)$ has constant multiplicity m_h. Here it is essential that D is simple (contains no exceptional point); in fact, $P_1(\varkappa_0)$ is not even defined if, for example, $\lambda_1(\varkappa_0) = \lambda_2(\varkappa_0)$ which may happen if $\varkappa_0$ is exceptional.

Let $\mathsf{M}_h(\varkappa) = P_h(\varkappa)\,\mathsf{X}$ be the (algebraic) eigenspace of $T(\varkappa)$ for the eigenvalue $\lambda_h(\varkappa)$. We have [see I-(5.34)]

(1.20) $$\mathsf{X} = \mathsf{M}_1(\varkappa) \oplus \cdots \oplus \mathsf{M}_s(\varkappa) ,$$
$$\dim \mathsf{M}_h(\varkappa) = m_h , \quad \sum_{j=1}^{s} m_h = N , \quad \varkappa \in \mathrm{D} .$$

The eigennilpotent $D_h(\varkappa)$ for the eigenvalue $\lambda_h(\varkappa)$ is also holomorphic for $\varkappa \in$ D, for

(1.21) $$D_h(\varkappa) = (T(\varkappa) - \lambda_h(\varkappa))\, P_h(\varkappa)$$

by I-(5.26).

There is a remarkable stability in the structure of the operators $P_h(\varkappa)$ and $D_h(\varkappa)$.

For each fixed h, dim $\mathsf{M}_h(\varkappa) = m_h$ is constant and the $\mathsf{M}_h(\varkappa)$ for different $\varkappa \in$ D are mutually isomorphic. Hence the $P_h(\varkappa)$ are mutually similar (see I-§ 4.6), i.e. there is a nonsingular operator $U_h(\varkappa)$ such that $U_h(\varkappa)^{-1} P_h(\varkappa) U_h(\varkappa) = P_h(\varkappa_0)$, where $\varkappa_0 \in$ D is fixed.

It is known that $U_h(\varkappa)$ can be chosen holomorphic in $\varkappa$, at least in a simple subdomain D. The proof is not very simple, and will be given later (see § 4).

The structure of the $D_h(\varkappa)$ is almost as stable. We have[1]

Theorem 1.7a. (stability of the Jordan canonical form) *For each fixed h, the $D_h(\varkappa)$ for different $\varkappa$ are mutually similar, except for $\varkappa$ in a certain isolated set. Furthermore, the similarity can be implemented by a meromorphic operator-function $V_h(\varkappa)$: $V_h(\varkappa)^{-1} D_h(\varkappa) V_h(\varkappa) = D_h(\varkappa_0)$, and $V_h(\varkappa)$* or *$V_h(\varkappa)^{-1}$ has poles at the exceptional points mentioned.*

The appearance of the (new kind of) exceptional points in Theorem 1.7a cannot in general be avoided. A simple example is given by $T(\varkappa) = \begin{pmatrix} 0 & \varkappa \\ 0 & 0 \end{pmatrix}$ of Example 1.1, c). $T(\varkappa)$ is itself nilpotent and coincides with the eigennilpotent $D(\varkappa)$ for the eigenvalue $\lambda(\varkappa) = 0$ with the eigenprojection $P(\varkappa) = 1$ (see Example 1.12, c) below). The general $V(\varkappa)$ that makes $V(\varkappa)^{-1} D(\varkappa) V(\varkappa)$ constant is of the form $V(\varkappa) = \begin{pmatrix} 1 & b \\ 0 & 1/\varkappa \end{pmatrix}$ (except for a scalar factor). It is impossible to eliminate a pole at $\varkappa = 0$ from both $V(\varkappa)$ and $V(\varkappa)^{-1}$ by adjusting a scalar factor for $V(\varkappa)$. In fact $D(0)$ is *not* similar to $D(\varkappa)$ with $\varkappa \neq 0$.

The proof of Theorem 1.7a is simple. We have only to repeat the construction of the Jordan canonical form sketched in I-§ 1.4 for a nilpotent matrix $D_h(\varkappa)$ with holomorphic matrix elements. To systematize this operation, it is convenient to introduce the vector space X_M *over the field* M *of all meromorphic functions in* D, and the related notion of matrices with elements in M. Since the construction mentioned above uses only linear operations, it is directly applicable to $D_h(\varkappa)$ regarded as a nilpotent matrix D_M with elements in M. Thus D_M can be brought into the form I-(3.30) by introducing a suitable new basis in X_M. This amounts to saying that there is a nonsingular operator $V_h(\varkappa)$, depending on $\varkappa$ *meromorphically*, such that $V_h(\varkappa)^{-1} D_h(\varkappa) V_h(\varkappa)$ is independent of $\varkappa$ and has a matrix of the special form I-(3.30). The poles of $V_h(\varkappa)$ or $V_h(\varkappa)^{-1}$ are the exceptional points mentioned in the theorem[2].

---

[1] These results were proved by Baumgärtel [5] (see also [[1]]).

[2] It can be shown, by a similar argument, that there is a *meromorphic* function $U_h(\varkappa)$ such that $U_h(\varkappa)^{-1} P_h(\varkappa) U_h(\varkappa) = P_h(\varkappa_0)$. This is not a satisfactory result, however, since it is known that $U_h(\varkappa)$ can be chosen *holomorphic*. Indeed one can eliminate poles from $U_h(\varkappa)$ and $U_h(\varkappa)^{-1}$ (see §§ 4.4, 4.9).

5. Singularities of the eigenprojections

Let us now consider the behavior of the eigenprojections $P_h(\varkappa)$ near an exceptional point, which we may again assume to be $\varkappa = 0$. As was shown above, each eigenvalue λ of T splits in general into several eigenvalues of $T(\varkappa)$ for $\varkappa \neq 0$, but the corresponding total projection is holomorphic at $\varkappa = 0$ [see (1.17)]. Take again the small disk D near $\varkappa = 0$ considered in par. 2; the eigenvalues $\lambda_h(\varkappa)$, the eigenprojections $P_h(\varkappa)$ and the eigennilpotents $D_h(\varkappa)$ are defined and holomorphic for $\varkappa \in$ D as shown above. When D is moved around $\varkappa = 0$ and brought to the initial position in the manner described in par. 2, each of the families $\{\lambda_h(\varkappa)\}$, $\{P_h(\varkappa)\}$ and $\{D_h(\varkappa)\}$ is subjected to a permutation by the analytic continuation. This permutation must be identical for the three families, as is seen from the following consideration.

The resolvent $R(\zeta, \varkappa)$ of $T(\varkappa)$ has the partial-fraction expression

$$(1.22) \quad R(\zeta, \varkappa) = -\sum_{h=1}^{s} \left[\frac{P_h(\varkappa)}{\zeta - \lambda_h(\varkappa)} + \frac{D_h(\varkappa)}{(\zeta - \lambda_h(\varkappa))^2} + \cdots + \frac{D_h(\varkappa)^{m_h - 1}}{(\zeta - \lambda_h(\varkappa))^{m_h}} \right]$$

[see I-(5.23)], where ζ is assumed to be somewhere distant from the spectrum of T so that $\zeta \in \mathrm{P}(T(\varkappa))$ for all $\varkappa$ considered. If $\lambda_1(\varkappa), \ldots, \lambda_p(\varkappa)$ constitute a cycle (see par. 2) of eigenvalues, the permutation mentioned above takes $\lambda_h(\varkappa)$ into $\lambda_{h+1}(\varkappa)$ for $1 \leqq h \leqq p - 1$ and $\lambda_p(\varkappa)$ into $\lambda_1(\varkappa)$ But as $R(\zeta, \varkappa)$ should be unchanged by the analytic continuation under consideration, the permutation must take $P_h(\varkappa)$ into $P_{h+1}(\varkappa)$ for $1 \leqq h \leqq p - 1$ and $P_p(\varkappa)$ into $P_1(\varkappa)$ [1]; the possibility that $P_h(\varkappa) = P_k(\varkappa)$ for some $h \neq k$ is excluded by the property $P_h(\varkappa) P_k(\varkappa) = \delta_{hk} P_h(\varkappa)$. Similar results hold for the eigennilpotents $D_h(\varkappa)$ by (1.21), except that some pair of the $D_h(\varkappa)$ may coincide [in fact all $D_h(\varkappa)$ can be zero).

We shall now show that $P_h(\varkappa)$ and $D_h(\varkappa)$ have at most algebraic singularities. Since $D_h(\varkappa)$ is given by (1.21), it suffices to prove this for $P_h(\varkappa)$. To this end we first note that

$$(1.23) \quad \|P_h(\varkappa)\| = \left\| \frac{1}{2\pi i} \int_{\Gamma_h(\varkappa)} R(\zeta, \varkappa)\, d\zeta \right\| \leqq \varrho_h(\varkappa) \max_{\zeta \in \Gamma_h(\varkappa)} \|R(\zeta, \varkappa)\|$$

where $\Gamma_h(\varkappa)$ is a circle enclosing $\lambda_h(\varkappa)$ but excluding all other $\lambda_k(\varkappa)$ and where $\varrho_h(\varkappa)$ denotes the radius of $\Gamma_h(\varkappa)$. On the other hand, we see from I-(4.12) that

$$(1.24) \quad \begin{aligned} \|R(\zeta, \varkappa)\| &= \|(T(\varkappa) - \zeta)^{-1}\| \leqq \\ &\leqq \gamma \|T(\varkappa) - \zeta\|^{N-1} / |\det(T(\varkappa) - \zeta)| \leqq \\ &\leqq \gamma (\|T(\varkappa)\| + |\zeta|)^{N-1} \Big/ \prod_{k=1}^{s} |\zeta - \lambda_k(\varkappa)|^{m_k}, \end{aligned}$$

[1] This is due to the uniqueness of the partial-fraction representation of $R(\zeta, \varkappa)$ as a function of ζ. A similar argument was used in I-§ 5.4.

where γ is a constant depending only on the norm employed. Hence

$$\|P_h(\varkappa)\| \leqq \gamma\, \varrho_h(\varkappa) \max_{\zeta \in \Gamma_h(\varkappa)} (\|T(\varkappa)\| + |\zeta|)^{N-1} \Big/ \prod_{k=1}^{s} |\zeta - \lambda_k(\varkappa)|^{m_k} . \tag{1.25}$$

Suppose that $\varkappa \to 0$, assuming again that $\varkappa = 0$ is an exceptional point. Then we have to choose the circle $\Gamma_h(\varkappa)$ smaller for smaller $|\varkappa|$ in order to ensure that it encloses $\lambda_h(\varkappa)$ but no other $\lambda_k(\varkappa)$, for the $\lambda_k(\varkappa)$ of the λ-group will approach $\lambda_h(\varkappa)$ indefinitely. But we know that the distances $|\lambda_h(\varkappa) - \lambda_k(\varkappa)|$ between these eigenvalues tend to zero for $\varkappa \to 0$ at most with some definite fractional order of $|\varkappa|$ because all $\lambda_k(\varkappa)$ have at most algebraic singularities at $\varkappa = 0$ [see (1.7)]. By choosing $\varrho_h(\varkappa) = |\varkappa|^\alpha$ with an appropriate $\alpha > 0$, we can therefore ensure that $\prod |\zeta - \lambda_k(\varkappa)|^{m_k} \geqq \gamma' |\varkappa|^{\alpha N}$ for $\zeta \in \Gamma_h(\varkappa)$ with some constant $\gamma' > 0$. Then we have

$$\|P_h(\varkappa)\| \leqq \text{const.}\ |\varkappa|^{-(N-1)\alpha} . \tag{1.26}$$

This shows that, when $P_h(\varkappa)$ is represented by a Laurent series in $\varkappa^{1/p}$, the principal part is finite.

These results may be summarized in

Theorem 1.8. *The eigenvalues $\lambda_h(\varkappa)$, the eigenprojections $P_h(\varkappa)$ and the eigennilpotents $D_h(\varkappa)$ of $T(\varkappa)$ are (branches of) analytic functions for $\varkappa \in \mathrm{D}_0$ with only algebraic singularities at some (but not necessarily all) exceptional points. $\lambda_h(\varkappa)$ and $P_h(\varkappa)$ have all branch points in common (including the order of the branch points), which may or may not be branch points for $D_h(\varkappa)$. If in particular $\lambda_h(\varkappa)$ is single-valued near an exceptional point $\varkappa = \varkappa_0$ (cycle of period 1), then $P_h(\varkappa)$ and $D_h(\varkappa)$ are also single-valued there.*

6. Remarks and examples

Although the $P_h(\varkappa)$ and $D_h(\varkappa)$ have algebraic singularities as well as the $\lambda_h(\varkappa)$, there are some important differences in their behavior at the singular points. Roughly speaking, $P_h(\varkappa)$ and $D_h(\varkappa)$ have stronger singularities than $\lambda_h(\varkappa)$.

We recall that these singular points are exceptional points, though the converse is not true. As we have already noted, the $\lambda_h(\varkappa)$ are continuous even at exceptional points and, therefore, have no poles. But $P_h(\varkappa)$ and $D_h(\varkappa)$ are in general undefined at exceptional points. In particular they may be single-valued and yet have a pole at an exceptional point (see Example 1.12 below).

Even more remarkable is the following theorem[1].

Theorem 1.9. *If $\varkappa = \varkappa_0$ is a branch point of $\lambda_h(\varkappa)$ (and therefore also of $P_h(\varkappa)$) of order $p - 1 \geqq 1$, then $P_h(\varkappa)$ has a pole there; that is, the*

[1] This theorem is due to Butler [1].

Laurent expansion of $P_h(\varkappa)$ *in powers of* $(\varkappa - \varkappa_0)^{1/p}$ *necessarily contains negative powers. In particular* $\|P_h(\varkappa)\| \to \infty$ *for* $\varkappa \to \varkappa_0$.

Proof. Suppose that this were not the case and let

$$P_h(\varkappa) = P_h + \varkappa^{1/p} P'_h + \cdots, \quad h = 1, \ldots, p,$$

be the Puiseux series for the $P_h(\varkappa)$ belonging to the cycle under consideration. Here we again assume for simplicity that $\varkappa_0 = 0$. When $\varkappa$ is subjected to a revolution around $\varkappa = 0$, $P_h(\varkappa)$ is changed into $P_{h+1}(\varkappa)$ for $1 \leqq h \leqq p-1$ and $P_p(\varkappa)$ into $P_1(\varkappa)$. Hence we must have $P_{h+1} = P_h$ for $1 \leqq h \leqq p-1$. On the other hand, the relation $P_h(\varkappa)\, P_{h+1}(\varkappa) = 0$ for $\varkappa \to 0$ gives $P_h\, P_{h+1} = 0$, and the idempotent character of $P_h(\varkappa)$ gives $P_h^2 = P_h$. Hence $P_h = P_h^2 = P_h\, P_{h+1} = 0$. But this contradicts the fact that $\dim P_h(\varkappa)\, \mathsf{X} = m_h > 0$, which implies that $\|P_h(\varkappa)\| \geqq 1$ (see Problem I-4.1).

As regards the order $p - 1$ of the branch point $\varkappa = \varkappa_0$ for $\lambda_h(\varkappa)$ or, equivalently, the period p of the cycle $\{\lambda_1(\varkappa), \ldots, \lambda_p(\varkappa)\}$, we have the following result. *An eigenvalue* λ *of* T *with multiplicity* m *does not give rise to a branch point of order larger than* $m - 1$. This is an obvious consequence of the fact that such an eigenvalue can never split into more than m eigenvalues [see (1.19)].

Theorem 1.10. *Let* X *be a unitary space. Let* $\varkappa_0 \in \mathrm{D}_0$ *(possibly an exceptional point) and let there exist a sequence* $\{\varkappa_n\}$ *converging to* $\varkappa_0$ *such that* $T(\varkappa_n)$ *is normal for* $n = 1, 2, \ldots$. *Then all the* $\lambda_h(\varkappa)$ *and* $P_h(\varkappa)$ *are holomorphic at* $\varkappa = \varkappa_0$, *and the* $D_h(\varkappa) = 0$ *identically.*

Proof. We have $\|P_h(\varkappa_n)\| = 1$ since $T(\varkappa_n)$ is normal [see I-(6.64)]. Thus $\varkappa = \varkappa_0$ is not a branch point for any $\lambda_h(\varkappa)$ by Theorem 1.9. Consequently the $\lambda_h(\varkappa)$ are holomorphic at $\varkappa = \varkappa_0$. Then the $P_h(\varkappa)$ are single-valued there and, since they cannot have a pole for the same reason as above, they must be holomorphic. Then the $D_h(\varkappa)$ vanish identically, since the holomorphic functions $D_h(\varkappa) = (T(\varkappa) - \lambda_h(\varkappa))\, P_h(\varkappa)$ vanish at $\varkappa = \varkappa_n \to \varkappa_0$.

Remark 1.11. In general the $P_h(\varkappa)$ and $D_h(\varkappa)$ are not defined at an exceptional point $\varkappa_0$. But they can have a removable singularity at $\varkappa_0$ as in Theorem 1.10. In such a case $P_h(\varkappa_0)$ and $D_h(\varkappa_0)$ are well-defined, but they need not be the eigenprojection and eigennilpotent for the eigenvalue $\lambda_h(\varkappa_0)$ of $T(\varkappa_0)$. If, for example, $\lambda_1(\varkappa_0) = \lambda_2(\varkappa_0) \neq \lambda_k(\varkappa_0)$, $k \geqq 3$, then $P_1(\varkappa_0) + P_2(\varkappa_0)$ (and not $P_1(\varkappa_0)$) is the eigenprojection for $\lambda_1(\varkappa_0)$. Again, the eigennilpotent for $\lambda_h(\varkappa_0)$ need not vanish even if $D_h(\varkappa) \equiv 0$, as is seen from Example 1.12 a), d), f) below.

Example 1.12. Consider the eigenprojections and eigennilpotents of $T(\varkappa)$ for the operators of Example 1.1.

a) The resolvent $R(\zeta, \varkappa)$ is given by (1.15), integration of which along small circles around $\lambda_{\pm}(\varkappa)$ gives by (1.16)

$$P_{\pm}(\varkappa) = \pm \frac{1}{2(1+\varkappa^2)^{1/2}} \begin{pmatrix} 1 \pm (1+\varkappa^2)^{1/2} & \varkappa \\ \varkappa & -1 \pm (1+\varkappa^2)^{1/2} \end{pmatrix}. \tag{1.27}$$

The reader is advised to verify the relations $P_{\pm}(\varkappa)^2 = P_{\pm}(\varkappa)$ and $P_{+}(\varkappa)\, P_{-}(\varkappa) = P_{-}(\varkappa)\, P_{+}(\varkappa) = 0$. The eigenprojections $P_{\pm}(\varkappa)$ are branches of a double-valued algebraic function with branch points $\varkappa = \pm i$. Since $s = N = 2$, $T(\varkappa)$ is simple and the eigennilpotents $D_{\pm}(\varkappa)$ are zero for $\varkappa \neq \pm i$. At the exceptional points $\varkappa = \pm i$, we have quite a different spectral representation of $T(\varkappa)$; there is a double eigenvalue 0, and the spectral representation of $T(\pm i)$ is

$$T(\pm i) = 0 + D_{\pm}, \tag{1.28}$$

that is, $T(\pm i)$ is itself the eigennilpotent.

b) Integration of the resolvent as in a) leads to the eigenprojections

$$P_1(\varkappa) = \frac{1}{2}\begin{pmatrix} 1 & 1 \\ 1 & 1 \end{pmatrix}, \quad P_2(\varkappa) = \frac{1}{2}\begin{pmatrix} 1 & -1 \\ -1 & 1 \end{pmatrix}. \tag{1.29}$$

for the eigenvalues $\lambda_1(\varkappa) = \varkappa$ and $\lambda_2(\varkappa) = -\varkappa$. Again we have $D_1(\varkappa) = D_2(\varkappa) = 0$ for $\varkappa \neq 0$. The exceptional point $\varkappa = 0$ is not a singular point for any of $\lambda_h(\varkappa)$, $P_h(\varkappa)$ or $D_h(\varkappa)$.

c) The eigenprojection and eigennilpotent for the unique eigenvalue $\lambda(\varkappa) = 0$ of $T(\varkappa)$ are given by $P(\varkappa) = 1$, $D(\varkappa) = T(\varkappa)$.

d) We have

$$P_{\pm}(\varkappa) = \frac{1}{2}\begin{pmatrix} 1 & \pm\varkappa^{-1/2} \\ \pm\varkappa^{1/2} & 1 \end{pmatrix}, \quad D_{\pm}(\varkappa) = 0, \quad \varkappa \neq 0, \tag{1.30}$$

for $\lambda_{\pm}(\varkappa) = \pm\varkappa^{1/2}$. The exceptional point $\varkappa = 0$ is a branch point for the eigenvalues and the eigenprojections. For $\varkappa = 0$, the eigenvalue is zero and the spectral representation is $T(0) = 0 + D$ with $D = T = T(0)$. The operator of this example resembles that of a), with the difference that there is only one exceptional point here.

e) We have

$$P_1(\varkappa) = \begin{pmatrix} 1 & \varkappa \\ 0 & 0 \end{pmatrix}, \quad P_2(\varkappa) = \begin{pmatrix} 0 & -\varkappa \\ 0 & 1 \end{pmatrix}, \quad D_h(\varkappa) = 0, \tag{1.31}$$

for $\lambda_1(\varkappa) = 1$ and $\lambda_2(\varkappa) = 0$. Everything is holomorphic for finite $\varkappa$ since there are no exceptional points. Note that the $P_h(\varkappa)$ are not holomorphic at $\varkappa = \infty$ whereas the $\lambda_h(\varkappa)$ are. This is a situation in a sense opposite to that of the following example.

f) The eigenprojections are

$$P_1(\varkappa) = \begin{pmatrix} 1 & \varkappa^{-1} \\ 0 & 0 \end{pmatrix}, \quad P_2(\varkappa) = \begin{pmatrix} 0 & -\varkappa^{-1} \\ 0 & 1 \end{pmatrix}, \quad \varkappa \neq 0, \tag{1.32}$$

for $\lambda_1(\varkappa) = \varkappa$ and $\lambda_2(\varkappa) = 0$. Note that the $P_h(\varkappa)$ have a pole at the exceptional point $\varkappa = 0$ notwithstanding that the $\lambda_h(\varkappa)$ are holomorphic there. The situation is reversed for $\varkappa = \infty$. At $\varkappa = 0$ the spectral representation is the same as in d).

7. The case of $T(\varkappa)$ linear in $\varkappa$

The foregoing general results are somewhat simplified in the case (1.1) in which $T(\varkappa)$ is linear in $\varkappa$. Then $T(\varkappa)$ is defined in the whole complex plane, which will be taken as the domain D_0. The coefficients of the characteristic equation (1.3) are polynomials in $\varkappa$ of degree not exceeding N. Hence the eigenvalues $\lambda_h(\varkappa)$ are branches of algebraic functions of

$\varkappa$. If the algebraic equation (1.3) is *irreducible*, there is only one N-valued algebraic function so that we have $s = N$. If (1.3) is *reducible*, the eigenvalues $\lambda_h(\varkappa)$ can be classified into several groups, each group corresponding to an algebraic function. If there happen to be identical ones among these algebraic functions, we have $s < N$ (permanent degeneracy)[1].

The algebraic functions $\lambda_h(\varkappa)$ have no pole at a finite value of $\varkappa$. At $\varkappa = \infty$ they have at most a pole of order 1; this is seen by writing (1.1) in the form

$$T(\varkappa) = \varkappa(T' + \varkappa^{-1} T)\,, \tag{1.33}$$

for the eigenvalues of $T' + \varkappa^{-1} T$ are continuous for $\varkappa^{-1} \to 0$. More precisely, these eigenvalues have the expansion $\mu_h + \beta_h(\varkappa^{-1})^{1/p} + \cdots$ (Puiseux series in $\varkappa^{-1}$), so that the eigenvalues of $T(\varkappa)$ have the form

$$\lambda_h(\varkappa) = \mu_h \varkappa + \beta_h \varkappa^{1-\frac{1}{p}} + \cdots, \quad \varkappa \to \infty\,. \tag{1.34}$$

Note that $P_h(\varkappa)$ or $D_h(\varkappa)$ may be holomorphic at $\varkappa = \infty$ even when $\lambda_h(\varkappa)$ is not [see Example 1.12, f)].

8. Summary

For convenience the main results obtained in the preceding paragraphs will be summarized here[2].

Let $T(\varkappa) \in \mathscr{B}(\mathsf{X})$ be a family holomorphic in a domain D_0 of the complex $\varkappa$-plane. The number s of eigenvalues of $T(\varkappa)$ is constant if $\varkappa$ is not one of the exceptional points, of which there are only a finite number in each compact subset of D_0. In each simple subdomain (simply connected subdomain containing no exceptional point) D of D_0, the eigenvalues of $T(\varkappa)$ can be expressed as s holomorphic functions $\lambda_h(\varkappa)$, $h = 1, \ldots, s$, the eigenvalue $\lambda_h(\varkappa)$ having constant multiplicity m_h. The $\lambda_h(\varkappa)$ are branches of one or several analytic functions on D_0, which have only algebraic singularities and which are everywhere continuous in D_0. [For simplicity these analytic functions will also be denoted by $\lambda_h(\varkappa)$.] An exceptional point $\varkappa_0$ is either a branch point of some of the $\lambda_h(\varkappa)$ or a regular point for all of them; in the latter case the values of some of the different $\lambda_h(\varkappa)$ coincide at $\varkappa = \varkappa_0$.

The eigenprojections $P_h(\varkappa)$ and the eigennilpotents $D_h(\varkappa)$ for the eigenvalue $\lambda_h(\varkappa)$ of $T(\varkappa)$ are branches of one or several analytic functions (again denoted by the same symbols) with only algebraic singularities. They are holomorphic in each subdomain D. For each h, $P_h(\varkappa)$ has constant dimension m_h, while $D_h(\varkappa)$ has constant Jordan canonical form except for $\varkappa$ in an isolated subset of D[3]. The analytic functions $P_h(\varkappa)$ and

[1] The results stated here are also true if $T(\varkappa)$ is a polynomial in $\varkappa$ of any degree.

[2] For more detailed and precise statement see BAUMGÄRTEL [1].

[3] For more details on the behavior of the $P_h(\varkappa)$ and the geometric eigenspaces, see § 4.7.

$\lambda_h(\varkappa)$ have common branch points of the same order, but $P_h(\varkappa)$ always has a pole at a branch point while $\lambda_h(\varkappa)$ is continuous there. $P_h(\varkappa)$ and $D_h(\varkappa)$ may have poles even at an exceptional point where $\lambda_h(\varkappa)$ is holomorphic.

If $\lambda_1(\varkappa), \ldots, \lambda_r(\varkappa)$ are the λ-group eigenvalues [the totality of the eigenvalues of $T(\varkappa)$ generated by splitting from a common eigenvalue λ of the unperturbed operator $T = T(0)$, $\varkappa = 0$ being assumed to be an exceptional point] and if $P_1(\varkappa), \ldots, P_r(\varkappa)$ are the associated eigenprojections, the total projection $P(\varkappa) = P_1(\varkappa) + \cdots + P_r(\varkappa)$ for this λ-group is holomorphic at $\varkappa = 0$. The total multiplicity $m_1 + \cdots + m_r$ for these eigenvalues is equal to the multiplicity m of the eigenvalue λ of T. The λ-group is further divided into several cycles $\{\lambda_1(\varkappa), \ldots, \lambda_p(\varkappa)\}$, $\{\lambda_{p+1}(\varkappa), \ldots\}$, $\ldots$, $\{\ldots\}$ and correspondingly for the eigenprojections. The elements of each cycle are permuted cyclically among themselves after analytic continuation when $\varkappa$ describes a small circle around $\varkappa = 0$. The sum of the eigenprojections in each cycle [for example $P_1(\varkappa) + \cdots + P_p(\varkappa)$] is single-valued at $\varkappa = 0$ but need not be holomorphic (it may have a pole).

§ 2. Perturbation series

1. The total projection for the λ-group

In the preceding section we were concerned with the general properties of the functions $\lambda_h(\varkappa)$, $P_h(\varkappa)$ and $D_h(\varkappa)$ representing respectively the eigenvalues, eigenprojections and eigennilpotents of an operator $T(\varkappa) \in \mathscr{B}(\mathsf{X})$ depending holomorphically on a complex parameter $\varkappa$. In the present section we shall construct explicitly the Taylor series (if they exist) for these functions at a given point $\varkappa$ which we may assume to be $\varkappa = 0$. Since the general case is too complicated to be dealt with completely, we shall be content with carrying out this program under certain simplifying assumptions. Furthermore, we shall give only formal series here; the convergence radii of the series and the error estimates will be considered in later sections[1].

We start from the given power series for $T(\varkappa)$:

$$T(\varkappa) = T + \varkappa\, T^{(1)} + \varkappa^2\, T^{(2)} + \cdots. \tag{2.1}$$

Let λ be one of the eigenvalues of the unperturbed operator $T = T(0)$ with (algebraic) multiplicity m, and let P and D be the associated eigen-

[1] The perturbation series have been studied extensively in quantum mechanics, starting with SCHRÖDINGER [1]. Any textbook on quantum mechanics has a chapter dealing with them (see e. g. KEMBLE [[1]], Chapter 11 or SCHIFF [[1]] Chapter 7). In most cases, however, the discussion is limited to selfadjoint (symmetric) operators $T(\varkappa)$ depending on a real parameter $\varkappa$. In this section we shall consider general nonsymmetric operators, assuming $0 < \dim \mathsf{X} = N < \infty$ as before.

projection and eigennilpotent. Thus (see I-§ 5.4)

$$(2.2)\quad TP = PT = PTP = \lambda P + D,\ \dim P = m,\ D^m = 0,\ PD = DP = D.$$

The eigenvalue λ will in general split into several eigenvalues of $T(\varkappa)$ for small $\varkappa \neq 0$ (the λ-group), see § 1.8. The total projection $P(\varkappa)$ for this λ-group is holomorphic at $\varkappa = 0$ [see (1.17)]

$$(2.3)\qquad P(\varkappa) = \sum_{n=0}^{\infty} \varkappa^n P^{(n)}\,,\quad P^{(0)} = P\,,$$

with $P^{(n)}$ given by (1.18). The subspace $\mathsf{M}(\varkappa) = P(\varkappa)\,\mathsf{X}$ is m-dimensional [see (1.19)] and invariant under $T(\varkappa)$. The λ-group eigenvalues of $T(\varkappa)$ are identical with all the eigenvalues of $T(\varkappa)$ in $\mathsf{M}(\varkappa)$ [that is, of the part of $T(\varkappa)$ in $\mathsf{M}(\varkappa)$]. In order to determine the λ-group eigenvalues, therefore, we have only to solve an eigenvalue problem in the subspace $\mathsf{M}(\varkappa)$, which is in general smaller than the whole space X.

The eigenvalue problem for $T(\varkappa)$ in $\mathsf{M}(\varkappa)$ is equivalent to the eigenvalue problem for the operator

$$(2.4)\qquad T_r(\varkappa) = T(\varkappa)\,P(\varkappa) = P(\varkappa)\,T(\varkappa) = P(\varkappa)\,T(\varkappa)\,P(\varkappa)\,,$$

see I-§ 5.1. Thus the λ-group eigenvalues of $T(\varkappa)$ are exactly those eigenvalues of $T_r(\varkappa)$ which are different from zero, provided that $|\lambda|$ is large enough to ensure that these eigenvalues do not vanish for the small $|\varkappa|$ under consideration[1]. The last condition does not restrict generality, for T could be replaced by $T + \alpha$ with a scalar α without changing the nature of the problem.

In any case it follows that

$$(2.5)\qquad \hat{\lambda}(\varkappa) = \frac{1}{m}\,\mathrm{tr}(T(\varkappa)\,P(\varkappa)) = \lambda + \frac{1}{m}\,\mathrm{tr}\big((T(\varkappa) - \lambda)\,P(\varkappa)\big)$$

is equal to the *weighted mean* of the λ-group eigenvalues of $T(\varkappa)$, where the weight is the multiplicity of each eigenvalue [see I-(5.40) and I-(3.25)]. If there is no splitting of λ so that the λ-group consists of a single eigenvalue $\lambda(\varkappa)$ with multiplicity m, we have

$$(2.6)\qquad \hat{\lambda}(\varkappa) = \lambda(\varkappa)\,;$$

in particular this is always true if $m = 1$. In such a case the eigenprojection associated with $\lambda(\varkappa)$ is exactly the total projection (2.3) and the eigennilpotent is given by [see I-(5.26)]

$$(2.7)\qquad D(\varkappa) = (T(\varkappa) - \lambda(\varkappa))\,P(\varkappa)\,.$$

These series give a complete solution to the eigenvalue problem for the λ-group in the case of no splitting, $\lambda(\varkappa)$, $P(\varkappa)$ and $D(\varkappa)$ being all holomorphic at $\varkappa = 0$.

---

[1] Note that $T_r(\varkappa)$ has the eigenvalue 0 with multiplicity $N - m$, with the eigenprojection $1 - P(\varkappa)$. Cf. also footnote [1] on p. 36.

Let us now consider the explicit form of the series (2.3) and (2.5) in terms of the coefficients $T^{(n)}$ of (2.1). It should be remarked at this point that we could use as well the coefficients of the series (2.4) instead of $T^{(n)}$, for the eigenvalues and eigenprojections are the same for $T(\varkappa)$ and $T_r(\varkappa)$ so far as concerns the λ-group[1].

The coefficients $P^{(n)}$ of (2.3) are given by (1.14) and (1.18). Thus

$$(2.8)\quad P^{(n)} = -\frac{1}{2\pi i} \sum_{\substack{\nu_1+\cdots+\nu_p=n\\ \nu_j\geqq 1}} (-1)^p \int_\Gamma R(\zeta)\, T^{(\nu_1)} R(\zeta)\, T^{(\nu_2)} \ldots T^{(\nu_p)} R(\zeta)\, d\zeta,$$

where Γ is a small, positively-oriented circle around λ. To evaluate the integral (2.8), we substitute for $R(\zeta)$ its Laurent expansion I-(5.18) at $\zeta = \lambda$, which we write for convenience in the form

$$(2.9)\qquad R(\zeta) = \sum_{n=-m}^{\infty} (\zeta - \lambda)^n\, S^{(n+1)}$$

with

$$(2.10)\qquad S^{(0)} = -P\,,\quad S^{(n)} = S^n\,,\quad S^{(-n)} = -D^n\,,\quad n \geqq 1\,.$$

Here $S = S(\lambda)$ is the value at $\zeta = \lambda$ of the reduced resolvent of T (see loc. cit.); thus we have by I-(5.19) and (5.26)

$$(2.11)\qquad SP = PS = 0\,,\quad (T-\lambda)\, S = S\,(T-\lambda) = 1 - P\,.$$

Substitution of (2.9) into the integrand of (2.8) gives a Laurent series in $\zeta - \lambda$, of which only the term with the power $(\zeta-\lambda)^{-1}$ contributes to the integral. The result is given by the *finite* sum

$$(2.12)\quad P^{(n)} = -\sum_{p=1}^{n} (-1)^p \sum_{\substack{\nu_1+\cdots+\nu_p=n\\ k_1+\cdots+k_{p+1}=p\\ \nu_j\geqq 1,\, k_j\geqq -m+1}} S^{(k_1)}\, T^{(\nu_1)}\, S^{(k_2)} \ldots S^{(k_p)}\, T^{(\nu_p)}\, S^{(k_{p+1})}$$

for $n \geqq 1$. For example

$$(2.13)\quad \begin{aligned} P^{(1)} &= \sum_{k_1+k_2=1} S^{(k_1)}\, T^{(1)}\, S^{(k_2)} \\ &= -D^{m-1}\, T^{(1)}\, S^m - \cdots - D\, T^{(1)}\, S^2 - P\, T^{(1)}\, S - S\, T^{(1)}\, P \\ &\quad - S^2\, T^{(1)}\, D - \cdots - S^m\, T^{(1)}\, D^{m-1}\,, \\ P^{(2)} &= \sum_{k_1+k_2=1} S^{(k_1)}\, T^{(2)}\, S^{(k_2)} - \sum_{k_1+k_2+k_3=2} S^{(k_1)}\, T^{(1)}\, S^{(k_2)}\, T^{(1)}\, S^{(k_3)}. \end{aligned}$$

If in particular λ is a semisimple eigenvalue of T (see I-§ 5.4), we have $D = 0$ and only nonnegative values of k_j contribute to the sum

[1] This remark will be useful later when we consider eigenvalue problems for unbounded operators in an infinite-dimensional space; it is then possible that the series (2.1) does not exist but (2.4) has a series expansion in $\varkappa$. See VII-§ 1.5.

(2.12). Thus we have, for example,

$$(2.14)\quad P^{(1)} = -PT^{(1)}S - ST^{(1)}P\,,$$

$$P^{(2)} = -PT^{(2)}S - ST^{(2)}P + PT^{(1)}ST^{(1)}S + ST^{(1)}PT^{(1)}S + \\ + ST^{(1)}ST^{(1)}P - PT^{(1)}PT^{(1)}S^2 - PT^{(1)}S^2T^{(1)}P - S^2T^{(1)}PT^{(1)}P\,,$$

$$P^{(3)} = -PT^{(3)}S - ST^{(3)}P + PT^{(1)}ST^{(2)}S + PT^{(2)}ST^{(1)}S + \\ + ST^{(1)}PT^{(2)}S + ST^{(2)}PT^{(1)}S + ST^{(1)}ST^{(2)}P + ST^{(2)}ST^{(1)}P - \\ - PT^{(1)}PT^{(2)}S^2 - PT^{(2)}PT^{(1)}S^2 - PT^{(1)}S^2T^{(2)}P - PT^{(2)}S^2T^{(1)}P - \\ - S^2T^{(1)}PT^{(2)}P - S^2T^{(2)}PT^{(1)}P - PT^{(1)}ST^{(1)}ST^{(1)}S - \\ - ST^{(1)}PT^{(1)}ST^{(1)}S - ST^{(1)}ST^{(1)}PT^{(1)}S - ST^{(1)}ST^{(1)}ST^{(1)}P + \\ + PT^{(1)}PT^{(1)}ST^{(1)}S^2 + PT^{(1)}PT^{(1)}S^2T^{(1)}S + PT^{(1)}ST^{(1)}PT^{(1)}S^2 + \\ + PT^{(1)}S^2T^{(1)}PT^{(1)}S + PT^{(1)}ST^{(1)}S^2T^{(1)}P + PT^{(1)}S^2T^{(1)}ST^{(1)}P + \\ + ST^{(1)}PT^{(1)}S^2T^{(1)}P + S^2T^{(1)}PT^{(1)}ST^{(1)}P + ST^{(1)}PT^{(1)}PT^{(1)}S^2 + \\ + S^2T^{(1)}PT^{(1)}PT^{(1)}S + ST^{(1)}S^2T^{(1)}PT^{(1)}P + S^2T^{(1)}ST^{(1)}PT^{(1)}P - \\ - PT^{(1)}PT^{(1)}PT^{(1)}S^3 - PT^{(1)}PT^{(1)}S^3T^{(1)}P - \\ - PT^{(1)}S^3T^{(1)}PT^{(1)}P - S^3T^{(1)}PT^{(1)}PT^{(1)}P\,.$$

2. The weighted mean of eigenvalues

We next consider the series (2.4) for $T_r(\varkappa) = T(\varkappa)\,P(\varkappa)$. For computation it is more convenient to consider the operator $(T(\varkappa) - \lambda)\,P(\varkappa)$ instead of $T_r(\varkappa)$ itself. We have from (1.16)

$$(2.15)\qquad (T(\varkappa) - \lambda)\,P(\varkappa) = -\frac{1}{2\pi i}\int_\Gamma (\zeta - \lambda)\,R(\zeta,\varkappa)\,d\zeta$$

since $(T(\varkappa) - \lambda)\,R(\zeta,\varkappa) = 1 + (\zeta - \lambda)\,R(\zeta,\varkappa)$ and the integral of 1 along Γ vanishes. Noting that $(T - \lambda)\,P = D$ by (2.2), we have

$$(2.16)\qquad (T(\varkappa) - \lambda)\,P(\varkappa) = D + \sum_{n=1}^{\infty} \varkappa^n\,\tilde{T}^{(n)}$$

with

$$(2.17)\quad \tilde{T}^{(n)} = -\frac{1}{2\pi i}\sum_{\substack{\nu_1+\cdots+\nu_p=n\\ \nu_j\ge 1}} (-1)^p \int_\Gamma R(\zeta)\,T^{(\nu_1)}\ldots T^{(\nu_p)}\,R(\zeta)\,(\zeta-\lambda)\,d\zeta$$

for $n \geqq 1$; this differs from (2.8) only by the factor $\zeta - \lambda$ in the integrand. Hence it follows that

$$(2.18)\qquad \tilde{T}^{(n)} = -\sum_{p=1}^{\infty} (-1)^p \sum_{\substack{\nu_1+\cdots+\nu_p=n\\ k_1+\cdots+k_{p+1}=p-1\\ \nu_j\ge 1,\ k_j\ge -m+1}}$$

with the same summand as in (2.12). For example

$$(2.19)\quad \tilde{T}^{(1)} = -D^{m-1}T^{(1)}S^{m-1} - \cdots - DT^{(1)}S + PT^{(1)}P - ST^{(1)}D \\ - \cdots - S^{m-1}T^{(1)}D^{m-1}\,.$$

Again these expressions are simplified when λ is a semisimple eigenvalue of $T(D=0)$; for example

$$(2.20)\quad \begin{aligned}\tilde{T}^{(1)} &= P T^{(1)} P\,,\\ \tilde{T}^{(2)} &= P T^{(2)} P - P T^{(1)} P T^{(1)} S - P T^{(1)} S T^{(1)} P - S T^{(1)} P T^{(1)} P,\\ \tilde{T}^{(3)} &= P T^{(3)} P - P T^{(1)} P T^{(2)} S - P T^{(2)} P T^{(1)} S -\end{aligned}$$

$$- P T^{(1)} S T^{(2)} P - P T^{(2)} S T^{(1)} P - S T^{(1)} P T^{(2)} P - S T^{(2)} P T^{(1)} P +$$
$$+ P T^{(1)} P T^{(1)} S T^{(1)} S + P T^{(1)} S T^{(1)} P T^{(1)} S +$$
$$+ P T^{(1)} S T^{(1)} S T^{(1)} P + S T^{(1)} P T^{(1)} P T^{(1)} S + S T^{(1)} P T^{(1)} S T^{(1)} P +$$
$$+ S T^{(1)} S T^{(1)} P T^{(1)} P - P T^{(1)} P T^{(1)} P T^{(1)} S^2 - P T^{(1)} P T^{(1)} S^2 T^{(1)} P -$$
$$- P T^{(1)} S^2 T^{(1)} P T^{(1)} P - S^2 T^{(1)} P T^{(1)} P T^{(1)} P\,.$$

The series for the weighted mean $\hat{\lambda}(\varkappa)$ of the λ-group eigenvalues is obtained from (2.5) and (2.16):

$$(2.21)\qquad \hat{\lambda}(\varkappa) = \lambda + \sum_{n=1}^{\infty} \varkappa^n \hat{\lambda}^{(n)}$$

where

$$(2.22)\qquad \hat{\lambda}^{(n)} = \frac{1}{m} \operatorname{tr} \tilde{T}^{(n)}\,, \quad n \geqq 1\,.$$

The substitution of (2.18) for $\tilde{T}^{(n)}$ will thus give the coefficients $\hat{\lambda}^{(n)}$.

But there is another expression for $\hat{\lambda}(\varkappa)$ which is more convenient for calculation, namely

$$(2.23)\qquad \hat{\lambda}(\varkappa) - \lambda = -\frac{1}{2\pi i m} \operatorname{tr} \int_\Gamma \log\left[1 + \left(\sum_{n=1}^{\infty} \varkappa^n T^{(n)}\right) R(\zeta)\right] d\zeta\,.$$

Here the logarithmic function $\log(1+A)$ is defined by

$$(2.24)\qquad \log(1+A) = \sum_{p=1}^{\infty} \frac{(-1)^{p-1}}{p} A^p$$

which is valid for $\|A\| < 1$. Note that (2.24) coincides with I-(5.57) for a special choice of the domain Δ (take as Δ a neighborhood of $\zeta = 1$ containing the eigenvalues of $1 + A$).

To prove (2.23), we start from (2.5) and (2.15), obtaining

$$(2.25)\qquad \hat{\lambda}(\varkappa) - \lambda = -\frac{1}{2\pi i m} \operatorname{tr} \int_\Gamma (\zeta - \lambda) R(\zeta, \varkappa)\, d\zeta\,.$$

Substitution for $R(\zeta, \varkappa)$ from (1.13) gives

$$(2.26)\qquad \hat{\lambda}(\varkappa) - \lambda = -\frac{1}{2\pi i m} \operatorname{tr} \int_\Gamma \sum_{p=1}^{\infty} (\zeta - \lambda) R(\zeta) (-A(\varkappa) R(\zeta))^p\, d\zeta\,;$$

note that the term for $p = 0$ in (2.26) vanishes because $\operatorname{tr} D = 0$ (see Problem I-3.11).

Now we have, in virtue of the relation $dR(\zeta)/d\zeta = R(\zeta)^2$ [see I-(5.8)],

$$(2.27)\qquad \begin{aligned}\frac{d}{d\zeta}(A(\varkappa) R(\zeta))^p &= \frac{d}{d\zeta}[A(\varkappa) R(\zeta) \ldots A(\varkappa) R(\zeta)]\\ &= A(\varkappa) R(\zeta) \ldots A(\varkappa) R(\zeta)^2 + \cdots + A(\varkappa) R(\zeta)^2 \ldots A(\varkappa) R(\zeta)\,.\end{aligned}$$

Application of the identity $\operatorname{tr} AB = \operatorname{tr} BA$ thus gives

$$\operatorname{tr} \frac{d}{d\zeta} (A(\varkappa) R(\zeta))^p = p \operatorname{tr} R(\zeta) (A(\varkappa) R(\zeta))^p , \tag{2.28}$$

and (2.26) becomes[1]

$$\hat{\lambda}(\varkappa) - \lambda = -\frac{1}{2\pi i m} \operatorname{tr} \int_\Gamma \sum_{p=1}^\infty \frac{1}{p} (\zeta - \lambda) \frac{d}{d\zeta} (-A(\varkappa) R(\zeta))^p \, d\zeta \tag{2.29}$$

$$= \frac{1}{2\pi i m} \operatorname{tr} \int_\Gamma \sum_{p=1}^\infty \frac{1}{p} (-A(\varkappa) R(\zeta))^p \, d\zeta \quad \text{(integration by parts)}$$

which is identical with (2.23) [recall the definition (1.10) of $A(\varkappa)$].

If the logarithmic function in (2.23) is expanded according to (2.24) and the result is arranged in powers of $\varkappa$, the coefficients in the series for $\hat{\lambda}(\varkappa)$ are seen to be given by

(2.30)
$$\hat{\lambda}^{(n)} = \frac{1}{2\pi i m} \operatorname{tr} \sum_{\nu_1 + \cdots + \nu_p = n} \frac{(-1)^p}{p} \int_\Gamma T^{(\nu_1)} R(\zeta) \ldots R(\zeta) T^{(\nu_p)} R(\zeta) \, d\zeta ; n \geqq 1 .$$

This can be treated as (2.8) and (2.17); the result is

$$\hat{\lambda}^{(n)} = \frac{1}{m} \sum_{p=1}^n \frac{(-1)^p}{p} \sum_{\substack{\nu_1 + \cdots + \nu_p = n \\ k_1 + \cdots + k_p = p-1}} \operatorname{tr} T^{(\nu_1)} S^{(k_1)} \ldots T^{(\nu_p)} S^{(k_p)} . \tag{2.31}$$

This formula is more convenient than (2.22), for the summation involved in (2.31) is simpler than in (2.18). For example,

$$\hat{\lambda}^{(1)} = \frac{1}{m} \operatorname{tr} T^{(1)} P , \tag{2.32}$$

$$\hat{\lambda}^{(2)} = \frac{1}{m} \left[\operatorname{tr} T^{(2)} P + \frac{1}{2} \sum_{k_1 + k_2 = 1} \operatorname{tr} T^{(1)} S^{(k_1)} T^{(1)} S^{(k_2)} \right]$$

$$= \frac{1}{m} [\operatorname{tr} T^{(2)} P - \operatorname{tr} (T^{(1)} S^m T^{(1)} D^{m-1} + \cdots + T^{(1)} S T^{(1)} P)] ,$$

where we have again used the identity $\operatorname{tr} AB = \operatorname{tr} BA$ [2].

These formulas are simplified when the eigenvalue λ is semisimple. Again making use of the identity mentioned, we thus obtain

$$\hat{\lambda}^{(1)} = \frac{1}{m} \operatorname{tr} T^{(1)} P , \tag{2.33}$$

$$\hat{\lambda}^{(2)} = \frac{1}{m} \operatorname{tr} [T^{(2)} P - T^{(1)} S T^{(1)} P]$$

$$\hat{\lambda}^{(3)} = \frac{1}{m} \operatorname{tr} [T^{(3)} P - T^{(1)} S T^{(2)} P - T^{(2)} S T^{(1)} P + $$
$$+ T^{(1)} S T^{(1)} S T^{(1)} P - T^{(1)} S^2 T^{(1)} P T^{(1)} P] ,$$

---

[1] The trace operation and the integration commute. The proof is similar to that of I-(4.30) and depends on the fact that tr is a linear functional on $\mathscr{B}(\mathsf{X})$.

[2] For example $(1/2) (\operatorname{tr} T^{(1)} S T^{(1)} P + \operatorname{tr} T^{(1)} P T^{(1)} S) = \operatorname{tr} T^{(1)} S T^{(1)} P$. Similar computations are made in the formulas (2.33).

$$\hat{\lambda}^{(4)} = \frac{1}{m} \operatorname{tr}[T^{(4)} P - T^{(1)} S T^{(3)} P - T^{(2)} S T^{(2)} P - T^{(3)} S T^{(1)} P$$
$$+ T^{(1)} S T^{(1)} S T^{(2)} P + T^{(1)} S T^{(2)} S T^{(1)} P + T^{(2)} S T^{(1)} S T^{(1)} P -$$
$$- T^{(1)} S^2 T^{(1)} P T^{(2)} P - T^{(1)} S^2 T^{(2)} P T^{(1)} P - T^{(2)} S^2 T^{(1)} P T^{(1)} P -$$
$$- T^{(1)} S T^{(1)} S T^{(1)} S T^{(1)} P + T^{(1)} S^2 T^{(1)} S T^{(1)} P T^{(1)} P +$$
$$+ T^{(1)} S T^{(1)} S^2 T^{(1)} P T^{(1)} P + T^{(1)} S^2 T^{(1)} P T^{(1)} S T^{(1)} P -$$
$$- T^{(1)} S^3 T^{(1)} P T^{(1)} P T^{(1)} P] .$$

Problem 2.1. If $T(\varkappa)$ is linear in $\varkappa$ ($T^{(n)} = 0$ for $n \geqq 2$), we have

$$\hat{\lambda}^{(n)} = \frac{1}{mn} \operatorname{tr}(T^{(1)} P^{(n-1)}), \quad n = 1, 2, 3, \ldots . \tag{2.34}$$

[hint: Compare (2.8) and (2.30).]

Remark 2.2. These expressions for the $\hat{\lambda}^{(n)}$ take more familiar (though more complicated) form if they are expressed in terms of bases chosen appropriately. For simplicity assume that the unperturbed operator T is diagonable. Let λ_h, P_h, $h = 1, 2, \ldots$, be the eigenvalues and eigenprojections of T different from the ones λ, P under consideration. Let $\{x_1, \ldots, x_m\}$ be a basis of $\mathsf{M} = \mathsf{R}(P)$ and let $\{x_{h1}, \ldots, x_{hm_h}\}$ be a basis of $\mathsf{M}_h = \mathsf{R}(P_h)$ for each h. The union of all these vectors x_j and x_{hj} forms a basis of X consisting of eigenvectors of T and adapted to the decomposition $\mathsf{X} = \mathsf{M} \oplus \mathsf{M}_1 \oplus \cdots$ of X. The adjoint basis of X^* is adapted to the corresponding decomposition $\mathsf{X}^* = \mathsf{M}^* \oplus \mathsf{M}_1^* \oplus \cdots$, where $\mathsf{M}^* = \mathsf{R}(P^*)$, $\mathsf{M}_1^* = \mathsf{R}(P_1^*)$, $\ldots$; it consists of a basis $\{e_1, \ldots, e_m\}$ of M^*, $\{e_{11}, \ldots, e_{1m_1}\}$ of M_1^*, etc. (see Problem I-3.19).

Now we have, for any $u \in \mathsf{X}$,

$$Pu = \sum_{j=1}^{m} (u, e_j)\, x_j , \quad P_h u = \sum_{j=1}^{m_h} (u, e_{hj})\, x_{hj} , \quad h = 1, 2, \ldots ,$$

and for any operator $A \in \mathscr{B}(\mathsf{X})$,

$$\operatorname{tr} A P = \sum_{j=1}^{m} (A x_j, e_j) , \quad \operatorname{tr} A P_h = \sum_{j=1}^{m_h} (A x_{hj}, e_{hj}) , \quad h = 1, 2, \ldots .$$

The operator S is given by I-(5.32) where the subscript h should be omitted. Hence

$$Su = \sum_k (\lambda_k - \lambda)^{-1} P_k u = \sum_{k,j} (\lambda_k - \lambda)^{-1} (u, e_{kj})\, x_{kj} .$$

Thus we obtain from (2.33) the following expressions for the $\hat{\lambda}^{(n)}$:

$$\hat{\lambda}^{(1)} = \frac{1}{m} \sum_j (T^{(1)} x_j, e_j) , \tag{2.35}$$

$$\hat{\lambda}^{(2)} = \frac{1}{m} \sum_j (T^{(2)} x_j, e_j) - \frac{1}{m} \sum_{i,j,k} (\lambda_k - \lambda)^{-1} (T^{(1)} x_i, e_{kj}) (T^{(1)} x_{kj}, e_i) .$$

Suppose, in particular, that the eigenvalue λ of T is simple: $m = 1$. Let φ be an eigenvector of T for this eigenvalue; then we can take $x_1 = \varphi$. Then $e_1 = \psi$ is an eigenvector of T^* for the eigenvalue $\bar{\lambda}$. We shall renumber the other eigenvectors x_{hj} of T in a simple sequence $\varphi_1, \varphi_2, \ldots$, with the corresponding eigenvalues $\mu_1, \mu_2, \ldots$ which are different from λ but not necessarily different from one another (repeated eigenvalues). Correspondingly, we write the e_{hj} in a simple sequence ψ_j so that $\{\psi, \psi_1, \psi_2, \ldots\}$ is the basis of X^* adjoint to the basis $\{\varphi, \varphi_1, \varphi_2, \ldots\}$ of X.

Then the above formulas can be written (note that $\hat{\lambda}^{(n)} = \lambda^{(n)}$ since there is no splitting)

$$\lambda^{(1)} = (T^{(1)}\varphi, \psi)\,, \tag{2.36}$$
$$\lambda^{(2)} = (T^{(2)}\varphi, \psi) - \sum_j (\mu_j - \lambda)^{-1} (T^{(1)}\varphi, \psi_j)\,(T^{(1)}\varphi_j, \psi)\,.$$

These are formulas familiar in the textbooks on quantum mechanics[1], except that here neither T nor the $T^{(n)}$ are assumed to be symmetric (Hermitian) and, therefore, we have a biorthogonal family of eigenvectors rather than an orthonormal family. (In the symmetric case we have $\psi = \varphi$, $\psi_j = \varphi_j$.)

3. The reduction process

If λ is a semisimple eigenvalue of T, $D = 0$ and (2.16) gives

$$\tilde{T}^{(1)}(\varkappa) = \frac{1}{\varkappa}(T(\varkappa) - \lambda)\, P(\varkappa) = \sum_{n=0}^{\infty} \varkappa^n \tilde{T}^{(n+1)}\,. \tag{2.37}$$

Since $\mathsf{M}(\varkappa) = \mathsf{R}(P(\varkappa))$ is invariant under $T(\varkappa)$, there is an obvious relationship between the parts of $T(\varkappa)$ and $\tilde{T}^{(1)}(\varkappa)$ in $\mathsf{M}(\varkappa)$. Thus the solution of the eigenvalue problem for $T(\varkappa)$ in $\mathsf{M}(\varkappa)$ reduces to the same problem for $\tilde{T}^{(1)}(\varkappa)$. Now (2.37) shows that $\tilde{T}^{(1)}(\varkappa)$ is holomorphic at $\varkappa = 0$, so that we can apply to it what has so far been proved for $T(\varkappa)$. This process of reducing the problem for $T(\varkappa)$ to the one for $\tilde{T}^{(1)}(\varkappa)$ will be called the *reduction process*. The "unperturbed operator" for this family $\tilde{T}^{(1)}(\varkappa)$ is [see (2.20)]

$$\tilde{T}^{(1)}(0) = \tilde{T}^{(1)} = P\,T^{(1)}\,P\,. \tag{2.38}$$

It follows that each eigenvalue of $\tilde{T}^{(1)}$ splits into several eigenvalues of $\tilde{T}^{(1)}(\varkappa)$ for small $|\varkappa|$. Let the eigenvalues of $\tilde{T}^{(1)}$ in the invariant subspace $\mathsf{M} = \mathsf{M}(0) = \mathsf{R}(P)$ be denoted by $\lambda_j^{(1)}$, $j = 1, 2, \ldots$ [the eigenvalue zero of $\tilde{T}^{(1)}$ in the complementary subspace $\mathsf{R}(1 - P)$ does not interest us]. The spectral representation of $\tilde{T}^{(1)}$ in M takes the form

$$\tilde{T}^{(1)} = P\,T^{(1)}\,P = \sum_j (\lambda_j^{(1)} P_j^{(1)} + D_j^{(1)})\,, \tag{2.39}$$
$$P = \sum_j P_j^{(1)}\,, \quad P_j^{(1)} P_k^{(1)} = \delta_{jk}\, P_j^{(1)}\,.$$

Suppose for the moment that all the $\lambda_j^{(1)}$ are different from zero. By perturbation each $\lambda_j^{(1)}$ will split into several eigenvalues (the $\lambda_j^{(1)}$-group) of $\tilde{T}^{(1)}(\varkappa)$, which are power series in $\varkappa^{1/p_j}$ with some $p_j \geqq 1$.[2] The corresponding eigenvalues of $T(\varkappa)$ have the form

$$\lambda + \varkappa\,\lambda_j^{(1)} + \varkappa^{1+\frac{1}{p_j}}\,\alpha_{jk} + \cdots\,, \quad k = 1, 2, \ldots\,. \tag{2.40}$$

[1] See Kemble ⟦1⟧ or Schiff ⟦1⟧, loc. cit.

[2] In general there are several cycles in the $\lambda_j^{(1)}$-group, but all eigenvalues of this group can formally be expressed as power series in $\varkappa^{1/p_j}$ for an appropriate common integer p_j.

If some $\lambda_j^{(1)}$ is zero, the associated eigenspace of $\tilde{T}^{(1)}$ includes the subspace $\mathsf{R}(1-P)$. But this inconvenience may be avoided by adding to $T(\varkappa)$ a term of the form $\alpha\varkappa$, which amounts to adding to $\tilde{T}^{(1)}(\varkappa)$ a term $\alpha P(\varkappa)$. This has only the effect of shifting the eigenvalues of $T^{(1)}(\varkappa)$ in $\mathsf{M}(\varkappa)$ [but not those in the complementary subspace $\mathsf{R}(1-P(\varkappa))$] by the amount α, leaving the eigenprojections and eigennilpotents unchanged. By choosing α appropriately the modified $\lambda_j^{(1)}$ can be made different from zero. Thus the assumption that $\lambda_j^{(1)} \neq 0$ does not affect the generality, and we shall assume this in the following whenever convenient.

The eigenvalues (2.40) of $T(\varkappa)$ for fixed λ and $\lambda_j^{(1)}$ will be said to form the $\lambda+\varkappa\lambda_j^{(1)}$-group. From (2.40) we see immediately that the following theorem holds.

Theorem 2.3. *If λ is a semisimple eigenvalue of the unperturbed operator T, each of the λ-group eigenvalues of $T(\varkappa)$ has the form* (2.40) *so that it belongs to some $\lambda+\varkappa\lambda_j^{(1)}$-group. These eigenvalues are continuously differentiable near $\varkappa=0$ (even when $\varkappa=0$ is a branch point). The total projection $P_j^{(1)}(\varkappa)$ (the sum of eigenprojections) for the $\lambda+\varkappa\lambda_j^{(1)}$-group and the weighted mean of this group are holomorphic at $\varkappa=0$.*

The last statement of the theorem follows from the fact that $P_j^{(1)}(\varkappa)$ is the total projection for the $\lambda_j^{(1)}$-group of the operator $\tilde{T}^{(1)}(\varkappa)$. The same is true for the weighted mean $\hat{\lambda}_j^{(1)}(\varkappa)$ of this $\lambda_j^{(1)}$-group.

The reduction process described above can further be applied to the eigenvalue $\lambda_j^{(1)}$ of $\tilde{T}^{(1)}$ if it is semisimple, with the result that the $\lambda_j^{(1)}$-group eigenvalues of $\tilde{T}^{(1)}(\varkappa)$ have the form $\lambda_j^{(1)}+\varkappa\lambda_{jk}^{(2)}+o(\varkappa)$. The corresponding eigenvalues of $T(\varkappa)$ have the form

$$\lambda+\varkappa\lambda_j^{(1)}+\varkappa^2\lambda_{jk}^{(2)}+o(\varkappa^2)\,. \tag{2.41}$$

These eigenvalues with fixed j, k form the $\lambda+\varkappa\lambda_j^{(1)}+\varkappa^2\lambda_{jk}^{(2)}$-group of $T(\varkappa)$. In this way we see that the reduction process can be continued, and the eigenvalues and eigenprojections of $T(\varkappa)$ can be expanded into formal power series in $\varkappa$, as long as the unperturbed eigenvalue is semisimple at each stage of the reduction process.

But it is not necessary to continue the reduction process indefinitely, even when this is possible. Since the splitting must end after a finite number, say n, of steps, the total projection and the weighted mean of the eigenvalues at the n-th stage will give the full expansion of the eigenprojection and the eigenvalue themselves, respectively.

Remark 2.4. But how can one know that there will be no splitting after the n-th stage? This is obvious if the total projection at that stage has dimension one. Otherwise there is no *general* criterion for it. In most applications, however, this problem can be solved by the following reducibility argument.

Suppose there is a set $\{A\}$ of operators such that $A\,T(\varkappa) = T(\varkappa)\,A$ for all $\varkappa$. Then A commutes with $R(\zeta, \varkappa)$ and hence with any eigenprojection of $T(\varkappa)$ [see I-(5.22)]. If there is any splitting of a semisimple eigenvalue λ of T, then, each $P_j^{(1)}(\varkappa)$ in Theorem 2.3 commutes with A and so does $P_j^{(1)} = P_j^{(1)}(0)$. Since $P_j^{(1)}$ is a proper subprojection (see I-§ 3.4) of P, we have the following result: *If λ is semisimple and P is irreducible in the sense that there is no subprojection of P which commutes with all A, then there is no splitting of λ at the first stage.* If it is known that the unperturbed eigenvalue in question is semisimple at every stage of the reduction process, then the irreducibility of P means that there is no splitting at all. Similarly, if the unperturbed eigenprojection becomes irreducible at some stage, there will be no further splitting.

4. Formulas for higher approximations

The series $P_j^{(1)}(\varkappa)$ for the total projection of the $\lambda + \varkappa\,\lambda_j^{(1)}$-group of $T(\varkappa)$ can be determined from the series (2.37) for $\tilde{T}^{(1)}(\varkappa)$ just as $P(\varkappa)$ was determined from $T(\varkappa)$. To this end we need the reduced resolvent for $\tilde{T}^{(1)}$, which corresponds to the reduced resolvent S of T used in the first stage. This operator will have the form

$$S_j^{(1)} - \frac{1}{\lambda_j^{(1)}}\,(1 - P) \tag{2.42}$$

where the second term comes from the part of $\tilde{T}^{(1)}$ in the subspace $(1 - P)\,\mathsf{X}$ in which $\tilde{T}^{(1)}$ is identically zero, and where

$$S_j^{(1)} = -\sum_{k \neq j}\left[\frac{P_k^{(1)}}{\lambda_j^{(1)} - \lambda_k^{(1)}} + \frac{D_k^{(1)}}{(\lambda_j^{(1)} - \lambda_k^{(1)})^2} + \cdots\right] \tag{2.43}$$

comes from the part of $\tilde{T}^{(1)}$ in $\mathsf{M} = P\mathsf{X}$ [see I-(5.32)]. We note that

$$S_j^{(1)} P = P\,S_j^{(1)} = S_j^{(1)}\,, \quad S_j^{(1)}\,P_j^{(1)} = P_j^{(1)}\,S_j^{(1)} = 0\,. \tag{2.44}$$

Application of the results of § 1.1 now gives

$$P_j^{(1)}(\varkappa) = P_j^{(1)} + \varkappa\,P_j^{(11)} + \varkappa^2\,P_j^{(12)} + \cdots \tag{2.45}$$

where the coefficients are calculated by (2.12), in which the $T^{(\nu)}$ are to be replaced by $\tilde{T}^{(\nu+1)}$ given by (2.18) and $S^{(k)}$ by $(S_j^{(1)} - (1 - P)/\lambda_j^{(1)})^k$ for $k \geqq 1$, by $-P_j^{(1)}$ for $k = 0$ and by $-(D_j^{(1)})^{-k}$ for $k \leqq -1$. If, for example, $\lambda_j^{(1)}$ is semisimple ($D_j^{(1)} = 0$), we have by (2.14)

$$P_j^{(11)} = -P_j^{(1)}\,\tilde{T}^{(2)}\left(S_j^{(1)} - \frac{1}{\lambda_j^{(1)}}\,(1 - P)\right) + (\text{inv}) \tag{2.46}$$

where (inv) means an expression obtained from the foregoing one by inverting the order of the factors in each term. Substitution of $\tilde{T}^{(2)}$

from (2.20) gives

$$P_j^{(11)} = -P_j^{(1)} T^{(2)} S_j^{(1)} - \frac{1}{\lambda_j^{(1)}} P_j^{(1)} T^{(1)} P T^{(1)} S + P_j^{(1)} T^{(1)} S T^{(1)} S_j^{(1)} + (\text{inv}).$$

But we have $P_j^{(1)} T^{(1)} P = P_j^{(1)} T^{(1)} P_j^{(1)} = \lambda_j^{(1)} P_j^{(1)}$ because $\lambda_j^{(1)}$ is a semisimple eigenvalue of $P T^{(1)} P$. Hence

$$\text{(2.47)} \quad P_j^{(11)} = -P_j^{(1)} T^{(2)} S_j^{(1)} - P_j^{(1)} T^{(1)} S + P_j^{(1)} T^{(1)} S T^{(1)} S_j^{(1)} + (\text{inv}).$$

Note that the final result does not contain $\lambda_j^{(1)}$ explicitly. Similarly $P_j^{(12)}$ can be calculated, though the expression will be rather complicated.

The weighted mean $\hat{\lambda}_j^{(1)}(\varkappa)$ for the $\lambda_j^{(1)}$-group eigenvalues of $\tilde{T}^{(1)}(\varkappa)$ is given by

$$\text{(2.48)} \qquad \hat{\lambda}_j^{(1)}(\varkappa) = \lambda_j^{(1)} + \varkappa\, \hat{\lambda}_j^{(12)} + \varkappa^2\, \hat{\lambda}_j^{(13)} + \cdots,$$

where the coefficients $\hat{\lambda}_j^{(1n)}$ are obtained from (2.31) by replacing m, P, S, $T^{(\nu)}$ by $m_j^{(1)} = \dim P_j^{(1)} \mathsf{X}$, $P_j^{(1)}$, (2.42) and $\tilde{T}^{(\nu+1)}$ respectively. For example, assuming that $\lambda_j^{(1)}$ is semisimple,

$$\text{(2.49)} \qquad \begin{aligned} \hat{\lambda}_j^{(12)} &= \frac{1}{m_j^{(1)}} \operatorname{tr} \tilde{T}^{(2)} P_j^{(1)} \\ &= \frac{1}{m_j^{(1)}} \operatorname{tr} [T^{(2)} P_j^{(1)} - T^{(1)} S T^{(1)} P_j^{(1)}], \\ \hat{\lambda}_j^{(13)} &= \frac{1}{m_j^{(1)}} \operatorname{tr} \Big[\tilde{T}^{(3)} P_j^{(1)} - \tilde{T}^{(2)} \Big(S_j^{(1)} - \frac{1}{\lambda_j^{(1)}} (1 - P)\Big) \tilde{T}^{(2)} P_j^{(1)}\Big] \\ &= \frac{1}{m_j^{(1)}} \operatorname{tr} [T^{(3)} P_j^{(1)} - T^{(1)} S T^{(2)} P_j^{(1)} - T^{(2)} S T^{(1)} P_j^{(1)} + \\ &\quad + T^{(1)} S T^{(1)} S T^{(1)} P_j^{(1)} - \lambda_j^{(1)} T^{(1)} S^2 T^{(1)} P_j^{(1)} - \\ &\quad - T^{(2)} S_j^{(1)} T^{(2)} P_j^{(1)} + T^{(1)} S T^{(1)} S_j^{(1)} T^{(2)} P_j^{(1)} + \\ &\quad + T^{(2)} S_j^{(1)} T^{(1)} S T^{(1)} P_j^{(1)} - T^{(1)} S T^{(1)} S_j^{(1)} T^{(1)} S T^{(1)} P_j^{(1)}]. \end{aligned}$$

Here we have again used the identity $\operatorname{tr} AB = \operatorname{tr} BA$ and the relations (2.44) and (2.11). The weighted mean of the $\lambda + \varkappa\, \lambda_j^{(1)}$-group eigenvalues of $T(\varkappa)$ is given by

$$\text{(2.50)} \qquad \begin{aligned} \hat{\lambda}_j(\varkappa) &= \lambda + \varkappa\, \hat{\lambda}_j^{(1)}(\varkappa) \\ &= \lambda + \varkappa\, \lambda_j^{(1)} + \varkappa^2\, \hat{\lambda}_j^{(12)} + \varkappa^3\, \hat{\lambda}_j^{(13)} + \cdots. \end{aligned}$$

If there is no splitting in the $\lambda + \varkappa\, \lambda_j^{(1)}$-group (that is, if this group consists of a single eigenvalue), (2.50) is exactly this eigenvalue. In particular this is the case if $m_j^{(1)} = 1$.

Remark 2.5. At first sight it might appear strange that the *third* order coefficient $\hat{\lambda}_j^{(13)}$ of (2.49) contains a term such as $-(1/m_j^{(1)}) \operatorname{tr} T^{(2)} S_j^{(1)} T^{(2)} P_j^{(1)}$ which is *quadratic in* $T^{(2)}$. But this does not involve any contradiction, as may be attested by the following example. Let $N = 2$ and let

$$T = 0\,, \quad T^{(1)} = \begin{pmatrix} 1 & 0 \\ 0 & -1 \end{pmatrix}, \quad T^{(2)} = \begin{pmatrix} 0 & \alpha \\ \alpha & 0 \end{pmatrix}, \quad T^{(n)} = 0 \quad \text{for } n \geqq 3\,.$$

The eigenvalues of $T(\varkappa)$ are

$$\pm \varkappa (1 + \alpha^2 \varkappa^2)^{1/2} = \pm \left(\varkappa + \frac{1}{2} \alpha^2 \varkappa^3 + \cdots \right),$$

in which the coefficient of the third order term is $\alpha^2/2$ and is indeed quadratic in α (that is, in $T^{(2)}$).

5. A theorem of MOTZKIN-TAUSSKY

As an application of Theorem 2.3, we shall prove some theorems due to MOTZKIN and TAUSSKY [1], [2].

Theorem 2.6. *Let the operator* $T(\varkappa) = T + \varkappa T'$ *be diagonable for every complex number* $\varkappa$. *Then all eigenvalues of* $T(\varkappa)$ *are linear in* $\varkappa$ *(that is, of the form* $\lambda_h + \varkappa\, \alpha_h$*), and the associated eigenprojections are entire functions of* $\varkappa$.

Proof. The eigenvalues $\lambda_h(\varkappa)$ of $T(\varkappa)$ are branches of algebraic functions of $\varkappa$ (see § 1.7). According to Theorem 2.3, on the other hand, the $\lambda_h(\varkappa)$ are continuously differentiable in $\varkappa$ at every (finite) value of $\varkappa$. Furthermore, we see from (1.34) that the $d\lambda_h(\varkappa)/d\varkappa$ are bounded at $\varkappa = \infty$. It follows that these derivatives must be constant (this is a simple consequence of the maximum principle for analytic functions[1]). This proves that the $\lambda_h(\varkappa)$ are linear in $\varkappa$.

Since $\lambda_h(\varkappa)$ has the form $\lambda_h + \varkappa\, \alpha_h$, the single eigenvalue $\lambda_h(\varkappa)$ constitutes the $\lambda_h + \varkappa\, \alpha_h$-group of $T(\varkappa)$. Thus the eigenprojection $P_h(\varkappa)$ associated with $\lambda_h(\varkappa)$ coincides with the total projection of this group and is therefore holomorphic at $\varkappa = 0$ (see Theorem 2.3). The same is true at every $\varkappa$ since $T(\varkappa)$ and $\lambda_h(\varkappa)$ are linear in $\varkappa$. Thus $P_h(\varkappa)$ is an entire function.

$P_h(\varkappa)$ may have a pole at $\varkappa = \infty$ [see Example 1.12, e)]. But if T' is also diagonable, $P_h(\varkappa)$ must be holomorphic even at $\varkappa = \infty$ because the eigenprojections of $T(\varkappa) = \varkappa(T' + \varkappa^{-1} T)$ coincide with those of $T' + \varkappa^{-1} T$, to which the above results apply at $\varkappa = \infty$. Hence each $P_h(\varkappa)$ is holomorphic everywhere including $\varkappa = \infty$ and so must be a constant by Liouville's theorem[2]. It follows that T and T' have common eigenprojections [namely $P_h(0) = P_h(\infty)$] and, since both are diagonable, they must commute. This gives

Theorem 2.7. *If* T' *is also diagonable in* Theorem 2.6, *then* T *and* T' *commute.*

---

[1] Since $\mu(\varkappa) = d\,\lambda_h(\varkappa)/d\varkappa$ is continuous everywhere (including $\varkappa = \infty$), $|\mu(\varkappa)|$ must take a maximum at some $\varkappa = \varkappa_0$ (possibly $\varkappa_0 = \infty$). Hence $\mu(\varkappa)$ must be constant by the maximum principle; see KNOPP ⟦1⟧, p. 84. [If $\varkappa_0$ is a branch point of order $p - 1$, apply the principle after the substitution $(\varkappa - \varkappa_0)^{1/p} = \varkappa'$; if $\varkappa_0 = \infty$, apply it after the substitution $\varkappa^{-1} = \varkappa'$.]

[2] See KNOPP ⟦1⟧, p. 112.

These theorems can be given a homogeneous form as follows.

Theorem 2.8. *Let A, $B \in \mathscr{B}(\mathsf{X})$ be such that their linear combination $\alpha A + \beta B$ is diagonable for all ratios $\alpha : \beta$ (including ∞) with possibly a single exception. Then all the eigenvalues of $\alpha A + \beta B$ have the form $\alpha \lambda_h + \beta \mu_h$ with λ_h, μ_h independent of α, β. If the said exception is excluded, then A and B commute.*

Remark 2.9. The above theorems are *global* in character in the sense that the diagonability of $T(\varkappa)$ $(\alpha A + \beta B)$ for *all* finite $\varkappa$ (*all* ratios $\alpha : \beta$ with a single exception) is essential. The fact that $T(\varkappa)$ is diagonable merely for all $\varkappa$ in some domain D of the $\varkappa$-plane does not even imply that the eigenvalues of $T(\varkappa)$ are holomorphic in D, as is seen from the following example.

Example 2.10. Let $N = 3$ and

$$T(\varkappa) = \begin{pmatrix} 0 & \varkappa & 0 \\ 0 & 0 & \varkappa \\ \varkappa & 0 & 1 \end{pmatrix} . \tag{2.51}$$

It is easy to see that $T(\varkappa)$ is diagonable for all $\varkappa$ with the exception of three values satisfying the equation $\varkappa^3 = -4/27$[1]. Thus $T(\varkappa)$ is diagonable for all $\varkappa$ in a certain neighborhood of $\varkappa = 0$. But the Puiseux series for the three eigenvalues of $T(\varkappa)$ for small $\varkappa$ have the forms

$$\pm \varkappa^{3/2} + \cdots, \quad 1 + \varkappa^3 + \cdots, \tag{2.52}$$

two of which are not holomorphic at $\varkappa = 0$.

Remark 2.11. Theorem 2.6 is not true in the case of infinite-dimensional spaces without further restrictions. Consider the differential operator

$$T(\varkappa) = -\frac{d^2}{dx^2} + x^2 + 2\varkappa x$$

regarded as a linear operator in the Hilbert space $\mathsf{L}^2(-\infty, +\infty)$ (such differential operators will be dealt with in detail in later chapters). $T(\varkappa)$ has the set of eigenvalues $\lambda_n(\varkappa)$ and the associated eigenfunctions $\varphi_n(x, \varkappa)$ given by

$$\lambda_n(\varkappa) = 2n - \varkappa^2, \qquad n = 0, 1, 2, \ldots,$$

$$\varphi_n(x, \varkappa) = \exp\left(-\frac{1}{2}x^2 - \varkappa x\right) H_n(x + \varkappa),$$

where the $H_n(x)$ are Hermite polynomials. The eigenfunctions φ_n form a complete set in the sense that every function of L^2 can be approximated with arbitrary precision by a linear combination of the φ_n. This is seen, for example, by noting that the set of functions of the form $\exp(-(x + \operatorname{Re}\varkappa)^2/2) \times$ (polynomial in x) is complete and that multiplication of a function by $\exp(-i x \operatorname{Im}\varkappa)$ is a unitary operator. Therefore, $T(\varkappa)$ may be regarded as *diagonable* for every finite $\varkappa$. Nevertheless, the $\lambda_n(\varkappa)$ are not linear in $\varkappa$.

6. The ranks of the coefficients of the perturbation series

The coefficients $P^{(n)}$ and $\tilde{T}^{(n)}$ of the series (2.3) and (2.16) have characteristic properties with respect to their ranks. Namely

---

[1] The characteristic equation for $T(\varkappa)$ is $\zeta^3 - \zeta^2 - \varkappa^3 = 0$. This cubic equation has 3 distinct roots so that $T(\varkappa)$ is diagonable, except when $\varkappa = 0$ or $\varkappa^3 = -4/27$. But $T(0)$ is obviously diagonable (it has already a diagonal matrix).

$$\text{(2.53)} \qquad \operatorname{rank} P^{(n)} \leqq (n+1)\, m\,, \quad \operatorname{rank} \tilde{T}^{(n)} \leqq (n+1)\, m\,, \qquad n = 1, 2, \ldots .$$

This follows directly from the following lemma.

Lemma 2.12. *Let $P(\varkappa) \in \mathscr{B}(\mathsf{X})$ and $A(\varkappa) \in \mathscr{B}(\mathsf{X})$ depend on $\varkappa$ holomorphically near $\varkappa = 0$ and let $P(\varkappa)$ be a projection for all $\varkappa$. Let $A(\varkappa)\, P(\varkappa)$ have the expansion*

$$\text{(2.54)} \qquad A(\varkappa)\, P(\varkappa) = \sum_{n=0}^{\infty} \varkappa^n B_n\,.$$

Then we have

$$\text{(2.55)} \qquad \operatorname{rank} B_n \leqq (n+1)\, m\,, \quad n = 0, 1, 2, \ldots,$$

where $m = \dim P(0)$. Similar results hold when the left member of (2.54) *is replaced by $P(\varkappa)\, A(\varkappa)$.*

Proof. Let

$$\text{(2.56)} \qquad P(\varkappa) = \sum_{n=0}^{\infty} \varkappa^n P_n\,.$$

The coefficients P_n satisfy recurrence formulas of the form

$$\text{(2.57)} \quad P_n = P_0 P_n + Q_{n1} P_0 P_{n-1} + \cdots + Q_{nn} P_0\,, \quad n = 0, 1, 2, \ldots,$$

where Q_{nk} is a certain polynomial in $P_0, P_1, \ldots, P_n$. (2.57) is proved by induction. The identity $P(\varkappa)^2 = P(\varkappa)$ implies that

$$\text{(2.58)} \qquad P_n = P_0 P_n + P_1 P_{n-1} + \cdots + P_n P_0\,,$$

which proves (2.57) for $n = 1$. If (2.57) is assumed to hold with n replaced by $1, 2, \ldots, n-1$, we have from (2.58)

$$\begin{aligned} P_n &= P_0 P_n + P_1 (P_0 P_{n-1} + Q_{n-1,1} P_0 P_{n-2} + \cdots + Q_{n-1,n-1} P_0) + \\ &+ P_2 (P_0 P_{n-2} + Q_{n-2,1} P_0 P_{n-3} + \cdots + Q_{n-2,n-2} P_0) + \cdots + P_n P_0 \\ &= P_0 P_n + P_1 P_0 P_{n-1} + (P_1 Q_{n-1,1} + P_2) P_0 P_{n-2} + \cdots + \\ &+ (P_1 Q_{n-1,n-1} + P_2 Q_{n-2,n-2} + \cdots + P_n) P_0\,, \end{aligned}$$

which is of the form (2.57). This completes the induction.

Now if $A(\varkappa) = \sum \varkappa^n A_n$ is the expansion of $A(\varkappa)$, we have from (2.54), (2.56) and (2.57)

$$\begin{aligned} \text{(2.59)} \quad B_n &= A_0 P_n + A_1 P_{n-1} + \cdots + A_n P_0 \\ &= A_0 P_0 P_n + (A_0 Q_{n1} + A_1) P_0 P_{n-1} + (A_0 Q_{n2} + A_1 Q_{n-1,1} + \\ &+ A_2) P_0 P_{n-2} + \cdots + (A_0 Q_{nn} + A_1 Q_{n-1,n-1} + \cdots + A_n) P_0\,. \end{aligned}$$

Thus B_n is the sum of $n + 1$ terms, each of which contains the factor P_0 and therefore has rank not exceeding $\operatorname{rank} P_0$ (see Problem I-3.4). This proves the required inequality (2.55). It is obvious how the above argument should be modified to prove the same results for $P(\varkappa)\, A(\varkappa)$ instead of $A(\varkappa)\, P(\varkappa)$.

§ 3. Convergence radii and error estimates

1. Simple estimates[1]

In the preceding sections we considered various power series in $\varkappa$ without giving explicit conditions for their convergence. In the present section we shall discuss such conditions.

We start from the expression of $R(\zeta, \varkappa)$ given by (1.13). This series is convergent for

$$\|A(\varkappa)\, R(\zeta)\| = \left\|\left(\sum_{n=1}^{\infty} \varkappa^n\, T^{(n)}\right) R(\zeta)\right\| < 1\,, \tag{3.1}$$

which is satisfied if

$$\sum_{n=1}^{\infty} |\varkappa|^n\, \|T^{(n)}\, R(\zeta)\| < 1\,. \tag{3.2}$$

Let $r(\zeta)$ be the value of $|\varkappa|$ such that the left member of (3.2) is equal to 1. Then (3.2) is satisfied for $|\varkappa| < r(\zeta)$.

Let the curve Γ be as in § 1.4. It is easily seen that (1.13) is uniformly convergent for $\zeta \in \Gamma$ if

$$|\varkappa| < r_0 = \min_{\zeta\in\Gamma} r(\zeta)\,, \tag{3.3}$$

so that the series (1.17) or (2.3) for the total projection $P(\varkappa)$ is convergent under the same condition (3.3). Thus r_0 is a lower bound for the convergence radius of the series for $P(\varkappa)$. Obviously r_0 is also a lower bound for the convergence radii of the series (2.21) for $\hat{\lambda}(\varkappa)$ and (2.37) for $\tilde{T}^{(1)}(\varkappa)$. Γ may be any simple, closed, rectifiable curve enclosing $\zeta = \lambda$ but excluding other eigenvalues of T, but we shall now assume that Γ is *convex*. It is convenient to choose Γ in such a way that r_0 turns out as large as possible.

To estimate the coefficients $\hat{\lambda}^{(n)}$ of (2.21), we use the fact that the λ-group eigenvalues of $T(\varkappa)$, and therefore also their weighted mean $\hat{\lambda}(\varkappa)$, lie inside Γ as long as (3.3) is satisfied[2]. On setting

$$\varrho = \max_{\zeta\in\Gamma} |\zeta - \lambda|\,, \tag{3.4}$$

we see that the function $\hat{\lambda}(\varkappa) - \lambda$ is holomorphic and bounded by ϱ for (3.3). It follows from Cauchy's inequality[3] for the Taylor coefficients that

$$|\hat{\lambda}^{(n)}| \leq \varrho\, r_0^{-n}\,, \quad n = 1, 2, \ldots\,. \tag{3.5}$$

---

[1] The following method, based on elementary results on function theory, is used by SZ.-NAGY [1], [2], T. KATO [1], [3], [6], SCHÄFKE [3], [4], [5].

[2] The convexity of Γ is used here.

[3] See KNOPP ⟦1⟧, p. 77.

Such estimates are useful in estimating the error incurred when the power series (2.21) is stopped after finitely many terms. Namely

$$\left|\hat{\lambda}(\varkappa) - \lambda - \sum_{p=1}^{n} \varkappa^p \hat{\lambda}^{(p)}\right| \leqq \sum_{p=n+1}^{\infty} |\varkappa|^p \, |\hat{\lambda}^{(p)}| \leqq \frac{\varrho \, |\varkappa|^{n+1}}{r_0^n (r_0 - |\varkappa|)} . \tag{3.6}$$

Example 3.1. Assume that

$$\|T^{(n)}\| \leqq a\, c^{n-1}, \quad n = 1, 2, \ldots . \tag{3.7}$$

for some nonnegative constants a, c. Such constants always exist since (1.2) is assumed convergent. Now (3.2) is satisfied if $a|\varkappa| (1 - c|\varkappa|)^{-1} \|R(\zeta)\| < 1$, that is, if $|\varkappa| < (a\|R(\zeta)\| + c)^{-1}$. Thus we have the following lower bound for the convergence radius

$$r_0 = \min_{\zeta \in \Gamma} (a\|R(\zeta)\| + c)^{-1} . \tag{3.8}$$

2. The method of majorizing series

Another method for estimating the coefficients and the convergence radii makes systematic use of majorizing series[1]. We introduce a *majorizing function (series)* $\Phi(\zeta - \lambda, \varkappa)$ for the function $A(\varkappa)\, R(\zeta)$ that appears in (1.13) and (2.29):

$$\begin{aligned} A(\varkappa)\, R(\zeta) &= \sum_{n=1}^{\infty} \varkappa^n T^{(n)} R(\zeta) \ll \Phi(\zeta - \lambda, \varkappa) \\ &\equiv \sum_{k=-\infty}^{\infty} \sum_{n=1}^{\infty} c_{kn} (\zeta - \lambda)^k \varkappa^n . \end{aligned} \tag{3.9}$$

By this we mean that each coefficient c_{kn} on the right is not smaller than the *norm* of the corresponding coefficient in the expansion of the left member in the double series in $\zeta - \lambda$ and $\varkappa$. Since $R(\zeta)$ has the Laurent expansion I-(5.18) with λ_h, P_h, D_h replaced by λ, P, D, respectively, this means that

$$\begin{aligned} &\|T^{(n)} D^k\| \leqq c_{-k-1,n} , \quad \|T^{(n)} P\| \leqq c_{-1,n} , \\ &\|T^{(n)} S^k\| \leqq c_{k-1,n} , \quad k > 0 . \end{aligned} \tag{3.10}$$

We assume that $c_{kn} = 0$ for $k < -m$ so that $\Phi(z, \varkappa)$ has only a pole at $z = 0$; this is allowed since $D^m = 0$.

In particular (3.9) implies

$$\left\|\sum_{n=1}^{\infty} \varkappa^n T^{(n)} R(\zeta)\right\| \leqq \Phi(|\zeta - \lambda|, |\varkappa|) . \tag{3.11}$$

Thus the series in (1.13) is convergent if $\Phi(|\zeta - \lambda|, |\varkappa|) < 1$. If we choose as Γ the circle $|\zeta - \lambda| = \varrho$, it follows that a lower bound r for the con-

[1] The use of majorizing series was begun by Rellich [4], and was further developed by Schröder [1]—[3]. Their methods are based on recurrence equations for the coefficients of the series, and differ from the function-theoretic approach used below.

vergence radius of the series for $P(\varkappa)$ as well as for $\hat{\lambda}(\varkappa)$ is given by the smallest positive root of the equation

$$\Phi(\varrho, r) = 1 . \tag{3.12}$$

It is convenient to choose ϱ so as to make this root r as large as possible.

Also Φ can be used to construct a majorizing series for $\hat{\lambda}(\varkappa) - \lambda$. Such a series is given by

$$\Psi(\varkappa) = \frac{1}{2\pi i} \int_{|\zeta-\lambda|=\varrho} -\log(1 - \Phi(\zeta - \lambda, \varkappa))\, d\zeta . \tag{3.13}$$

To see this we first note that, when the integrand in (2.29) is expanded into a power-Laurent series in $\varkappa$ and $\zeta - \lambda$ using (2.9), only those terms which contain at least one factor P or D contribute to the integral. Such terms are necessarily of rank $\leqq m$ and, therefore, their traces do not exceed m times their norms [see I-(5.42)]. Thus a majorizing series for $\hat{\lambda}(\varkappa) - \lambda$ is obtained if we drop the factor $1/m$ and the trace sign on the right of (2.29) and replace the coefficients of the expansion of the integrand by their norms. Since this majorizing function is in turn majorized by (3.13), the latter is seen to majorize $\hat{\lambda}(\varkappa) - \lambda$.

As is well known in function theory, (3.13) is equal to the sum of the zeros minus the sum of the poles of the function $1 - \Phi(z, \varkappa)$ contained in the interior of the circle $|z| = \varrho$ [1]. But the only pole of this function is $z = 0$ and does not contribute to the sum mentioned. In this way we have obtained

Theorem 3.2. *A majorizing series for $\hat{\lambda}(\varkappa) - \lambda$ is given by (the Taylor series representing) the sum of the zeros of the function $1 - \Phi(z, \varkappa)$ (as a function of z) in the neighborhood of $z = 0$ when $\varkappa \to 0$, multiple zeros being counted repeatedly. This majorizing series, and a fortiori the series for $P(\varkappa)$ and $\hat{\lambda}(\varkappa)$, converge for $|\varkappa| < r$, where r is the smallest positive root of (3.12); here ϱ is arbitrary as long as the circle $|\zeta - \lambda| = \varrho$ encloses no eigenvalues of T other than λ.*

Example 3.3. Consider the special case in which $T^{(n)} = 0$ for $n \geqq 2$ and λ is a semisimple eigenvalue of T $(D = 0)$. From (3.10) we see that we may take $c_{-1,1} = \|T^{(1)} P\|$, $c_{k1} \geqq \|T^{(1)} S^{k+1}\|$, $k \geqq 0$, all other c_{kn} being zero. For the choice of c_{k1}, we note that $S^{k+1} = S(S - \alpha P)^k$ for any α because $SP = PS = 0$. Thus we can take $c_{k1} = \|T^{(1)} S\| \, \|S - \alpha P\|^k$ and obtain

$$\Phi(z, \varkappa) = \varkappa \left(\frac{p}{z} + \frac{q}{1 - s z} \right) \tag{3.14}$$

as a majorizing series, where

$$p = \|T^{(1)} P\| , \quad q = \|T^{(1)} S\| , \quad s = \|S - \alpha P\| \quad \text{for any } \alpha . \tag{3.15}$$

---

[1] See KNOPP [1], p. 134; note that $\int \log f(z)\, dz = - \int f'(z)\, f(z)^{-1} z\, dz$ by integration by parts.

For small $|\varkappa|$ there is a unique zero $z = \Psi(\varkappa)$ of $1 - \Phi(z, \varkappa)$ in the neighborhood of $z = 0$. This $\Psi(\varkappa)$ is a majorizing series for $\hat{\lambda}(\varkappa) - \lambda$ by Theorem 3.2. A simple calculation gives

$$\begin{aligned}\Psi(\varkappa) &= p\varkappa + \frac{1}{2s}[1 - (ps+q)\varkappa - \Omega(\varkappa)] \\ &= p\varkappa + 2pq\varkappa^2[1 - (ps+q)\varkappa + \Omega(\varkappa)]^{-1} \\ &= p\varkappa + pq\varkappa^2 + \frac{2pq(ps+q)\varkappa^3 + 2p^2q^2s\varkappa^4}{1-(ps+q)\varkappa - 2pqs\varkappa^2 + \Omega(\varkappa)},\end{aligned} \tag{3.16}$$

where

$$\Omega(\varkappa) = \{[1 - (ps+q)\varkappa]^2 - 4pqs\varkappa^2\}^{1/2}. \tag{3.17}$$

Each coefficient of the power series of $\Psi(\varkappa)$ gives an upper bound for the corresponding coefficient of the series for $\hat{\lambda}(\varkappa) - \lambda$. Hence the remainder of this series after the n-th order term majorizes the corresponding remainder of $\hat{\lambda}(\varkappa)$. In this way we obtain from the second and third expressions of (3.16)

$$|\hat{\lambda}(\varkappa) - \lambda - \varkappa\hat{\lambda}^{(1)}| \leqq 2pq|\varkappa|^2/[1 - (ps+q)|\varkappa| + \Omega(|\varkappa|)], \tag{3.18}$$

$$|\hat{\lambda}(\varkappa) - \lambda - \varkappa\hat{\lambda}^{(1)} - \varkappa^2\hat{\lambda}^{(2)}| \leqq \frac{2pq|\varkappa|^3(ps+q+pqs|\varkappa|)}{1-(ps+q)|\varkappa| - 2pqs|\varkappa|^2 + \Omega(|\varkappa|)}. \tag{3.19}$$

Substitution of (3.14) into (3.12) gives a lower bound r for the convergence radii for the series of $P(\varkappa)$ and $\hat{\lambda}(\varkappa)$. The choice of

$$\varrho = p^{1/2}s^{-1/2}[(ps)^{1/2} + q^{1/2}]^{-1} \tag{3.20}$$

gives the best value[1]

$$r = [(ps)^{1/2} + q^{1/2}]^{-2}. \tag{3.21}$$

Note that the choice of (3.20) is permitted because $\varrho \leqq s^{-1} = \|S - \alpha P\|^{-1} \leqq d$, where d is the *isolation distance* of the eigenvalue λ of T (the distance of λ from other eigenvalues λ_k of T). In fact, I-(5.32) implies $(S - \alpha P)u = -(\lambda - \lambda_k)^{-1}u$ if $u = P_k u$ (note that $P_j P_k = 0$, $j \neq k$, $PP_k = 0$). Hence $\|S - \alpha P\| \geqq |\lambda - \lambda_k|^{-1}$ for all k.

Problem 3.4. In Example 3.3 we have

$$|\hat{\lambda}^{(1)}| \leqq p, \quad |\hat{\lambda}^{(2)}| \leqq pq, \quad |\hat{\lambda}^{(3)}| \leqq pq(ps+q), \ \ldots. \tag{3.22}$$

Remark 3.5. The series for $P(\varkappa)$ can also be estimated by using the majorizing function Φ. In virtue of (1.16), (1.13) and (3.9), we have

$$P(\varkappa) \ll \frac{1}{2\pi i}\int_{|\zeta-\lambda|=\varrho} \Phi_1(\zeta-\lambda)(1 - \Phi(\zeta-\lambda,\varkappa))^{-1}d\zeta \tag{3.23}$$

where $\Phi_1(\zeta - \lambda)$ is a majorizing series for $R(\zeta)$. The right member can be calculated by the method of residues if Φ and Φ_1 are given explicitly as in Example 3.3.

3. Estimates on eigenvectors

It is often required to calculate eigenvectors rather than eigenprojections. Since the eigenvectors are not uniquely determined, however, there are no definite formulas for the eigenvectors of $T(\varkappa)$ as

[1] This is seen also from (3.16), which has an expansion convergent for $|\varkappa| < r$ with r given by (3.21).

functions of $\varkappa$. If we assume for simplicity that $m = 1$ (so that $D = 0$), a convenient form of the eigenvector $\varphi(\varkappa)$ of $T(\varkappa)$ corresponding to the eigenvalue $\lambda(\varkappa)$ is given by

$$\varphi(\varkappa) = (P(\varkappa)\varphi, \psi)^{-1} P(\varkappa)\varphi , \tag{3.24}$$

where φ is an unperturbed eigenvector of T for the eigenvalue λ and ψ is an eigenvector of T^* for the eigenvalue $\bar{\lambda}$ normalized by $(\varphi, \psi) = 1$. Thus

$$P\varphi = \varphi , \quad P^*\psi = \psi , \quad (\varphi, \psi) = 1 . \tag{3.25}$$

That (3.24) is an eigenvector of $T(\varkappa)$ is obvious since $P(\varkappa)\mathsf{X}$ is one-dimensional. The choice of the factor in (3.24) is equivalent to each of the following normalization conditions:

$$(\varphi(\varkappa), \psi) = 1 , \quad (\varphi(\varkappa) - \varphi, \psi) = 0 , \quad P(\varphi(\varkappa) - \varphi) = 0 . \tag{3.26}$$

The relation $(T(\varkappa) - \lambda(\varkappa))\varphi(\varkappa) = 0$ can be written

$$(T - \lambda)(\varphi(\varkappa) - \varphi) + (A(\varkappa) - \lambda(\varkappa) + \lambda)\varphi(\varkappa) = 0 , \tag{3.27}$$

where $A(\varkappa) = T(\varkappa) - T$ [note that $(T - \lambda)\varphi = 0$]. Multiplying (3.27) from the left by S and using $S(T - \lambda) = 1 - P$ and (3.26), we have

$$\varphi(\varkappa) - \varphi + S[A(\varkappa) - \lambda(\varkappa) + \lambda]\varphi(\varkappa) = 0 . \tag{3.28}$$

Noting further that $S\varphi = 0$, we obtain [write $\varphi(\varkappa) = \varphi(\varkappa) - \varphi + \varphi$ in the last term of (3.28)]

$$\begin{aligned} \varphi(\varkappa) - \varphi &= -(1 + S[A(\varkappa) - \lambda(\varkappa) + \lambda])^{-1} S A(\varkappa)\varphi \\ &= -S[1 + A(\varkappa)S - (\lambda(\varkappa) - \lambda)S_\alpha]^{-1} A(\varkappa)\varphi \end{aligned} \tag{3.29}$$

if $\varkappa$ is sufficiently small, where $S_\alpha = S - \alpha P$ and α is arbitrary. This is a convenient formula for calculating an eigenvector.

In particular (3.29) gives the following majorizing series for $\varphi(\varkappa) - \varphi$:

$$\varphi(\varkappa) - \varphi \ll \|S\| (1 - \Phi_2(\varkappa) - \|S_\alpha\| \Psi(\varkappa))^{-1} \Phi_3(\varkappa) , \tag{3.30}$$

where $\Phi_2(\varkappa)$ and $\Phi_3(\varkappa)$ are majorizing series for $A(\varkappa)S$ and $A(\varkappa)\varphi$[1] respectively [note that $\Psi(\varkappa)$ is a majorizing series for $\lambda(\varkappa) - \lambda$]. The eigenvector $\varphi(\varkappa)$ is useful if $|\varkappa|$ is so small that the right member of (3.30) is smaller than $\|\varphi\|$, for then $\varphi(\varkappa)$ is certainly not zero.

Multiplication of (3.29) from the left by $T - \lambda$ gives

$$(T - \lambda)\varphi(\varkappa) = -(1 - P)[1 + A(\varkappa)S - (\lambda(\varkappa) - \lambda)S_\alpha]^{-1} A(\varkappa)\varphi \tag{3.31}$$

and hence

$$(T - \lambda)\varphi(\varkappa) \ll (1 - \Phi_2(\varkappa) - \|S_\alpha\| \Psi(\varkappa))^{-1} \Phi_3(\varkappa) . \tag{3.32}$$

---

[1] A majorizing series (function) for a vector-valued function can be defined in the same way as for an operator-valued function.

Example 3.6. If $T^{(n)} = 0$ for $n \geqq 2$, we have $A(\varkappa) = \varkappa T^{(1)}$ so that we may take

$$\Phi_2(\varkappa) = \varkappa \| T^{(1)} S \| = \varkappa q \,, \quad \Phi_3(\varkappa) = \varkappa \| T^{(1)} \varphi \| \,. \tag{3.33}$$

Thus (3.30) and (3.32), after substitution of (3.16), give ($s_0 = \|S\|$)

$$\varphi(\varkappa) - \varphi \ll \frac{2 \varkappa s_0 \| T^{(1)} \varphi \|}{1 - (p s + q) \varkappa + \{(1 - (p s + q) \varkappa)^2 - 4 p q s \varkappa^2\}^{1/2}} \,, \tag{3.34}$$

$$(T - \lambda)\, \varphi(\varkappa) \ll \text{[the right member of (3.34) with the factor } s_0 \text{ omitted]} \,. \tag{3.35}$$

Problem 3.7. Under the assumptions of Example 3.6, we have for $|\varkappa| < r$ [r is given by (3.21)]

$$\varphi(\varkappa) = \varphi - \varkappa S T^{(1)} \varphi + \varkappa^2 S (T^{(1)} - \lambda^{(1)}) S T^{(1)} \varphi - \cdots \,, \tag{3.36}$$

$$\varphi(\varkappa) - \varphi \ll s_0 \varkappa (1 + (p s + q) \varkappa + \cdots) \| T^{(1)} \varphi \| \,, \tag{3.37}$$

$$\| \varphi(\varkappa) - \varphi \| \leqq |\varkappa| \frac{s_0}{(p q s)^{1/2}} ((p s)^{1/2} + q^{1/2})^2 \| T^{(1)} \varphi \| \,, \tag{3.38}$$

$$\| \varphi(\varkappa) - \varphi + \varkappa S T^{(1)} \varphi \| \leqq |\varkappa|^2 \frac{s_0}{(p q s)^{1/2}} ((p s)^{1/2} + q^{1/2})^2 (p s + q + (p q s)^{1/2}) \,, \tag{3.39}$$

$$\| (T - \lambda)(\varphi(\varkappa) - \varphi) \| \leqq s_0^{-1} \text{ [right member of (3.38)]} \,, \tag{3.40}$$

$$\| (T - \lambda)(\varphi(\varkappa) - \varphi + \varkappa S T^{(1)} \varphi) \| \leqq s_0^{-1} \text{ [right member of (3.39)]} \,. \tag{3.41}$$

[hint for (3.38) and (3.39): Set $\varkappa = r$ after taking out the factor $|\varkappa|$ and $|\varkappa|^2$, respectively, from the majorizing series.]

4. Further error estimates

In view of the practical importance of the estimates for the remainder when the series of $\hat{\lambda}(\varkappa)$ is stopped after finitely many terms, we shall give other estimates of the coefficients $\hat{\lambda}^{(n)}$ than those given by (3.5) or by the majorizing series (3.13).

We write the integral expression (2.30) of $\hat{\lambda}^{(n)}$ in the following form:

$$\hat{\lambda}^{(n)} = \frac{1}{2\pi i m} \sum_{\nu_1 + \cdots + \nu_p = n} \frac{(-1)^p}{p} \int_\Gamma \operatorname{tr} [T^{(\nu_1)} R(\zeta) \ldots T^{(\nu_p)} R(\zeta) - T^{(\nu_1)} S(\zeta) \ldots T^{(\nu_p)} S(\zeta)] \, d\zeta \,. \tag{3.42}$$

Here $S(\zeta)$ is the reduced resolvent of T with respect to the eigenvalue λ (see I-§ 5.3), that is,

$$R(\zeta) = R_0(\zeta) + S(\zeta) \,, \quad R_0(\zeta) = P R(\zeta) = R(\zeta) P \,, \tag{3.43}$$

is the decomposition of $R(\zeta)$ into the principal part and the holomorphic part at the pole $\zeta = \lambda$. Note that the second term in the [ ] of (3.42) is holomorphic and does not contribute to the integral. Now this expression in [ ] is equal to

$$T^{(\nu_1)} R_0(\zeta) T^{(\nu_2)} R(\zeta) \ldots T^{(\nu_p)} R(\zeta) + T^{(\nu_1)} S(\zeta) T^{(\nu_2)} R_0(\zeta) \ldots T^{(\nu_p)} R(\zeta) + \cdots + T^{(\nu_1)} S(\zeta) \ldots T^{(\nu_{p-1})} S(\zeta) T^{(\nu_p)} R_0(\zeta) \,, \tag{3.44}$$

each term of which contains one factor $R_0(\zeta) = P R(\zeta)$. Since this factor has rank $\leqq m$, the same is true of each term of (3.44) and, consequently, the rank of (3.44) does not exceed $\min(p\,m, N)$, where $N = \dim \mathsf{X}$. Thus the trace in (3.42) is majorized in absolute value by $\min(p\,m, N)$ times the norm of this expression [see I-(5.42)]. This leads to the estimate

$$(3.45)\quad |\hat{\lambda}^{(n)}| \leq \frac{1}{2\pi} \sum_{\nu_1+\cdots+\nu_p=n} \min\left(1, \frac{N}{p\,m}\right) \int_\Gamma \| T^{(\nu_1)} R(\zeta) \ldots T^{(\nu_p)} R(\zeta) - T^{(\nu_1)} S(\zeta) \ldots T^{(\nu_p)} S(\zeta) \| \, |d\zeta| .$$

A somewhat different estimate can be obtained in the special case in which $T^{(n)} = 0$ for $n \geqq 2$. In this case we have (2.34), where $P^{(n-1)}$ has rank $\leqq \min(n\,m, N)$ by (2.53). Hence we have, again estimating the trace by $\min(n\,m, N)$ times the norm,

$$(3.46)\qquad |\hat{\lambda}^{(n)}| \leqq \min\left(1, \frac{N}{n\,m}\right) \| T^{(1)} P^{(n-1)} \| , \quad n = 1, 2, \ldots .$$

On the other hand (2.8) gives

$$(3.47)\quad \| T^{(1)} P^{(n-1)} \| \leqq \frac{1}{2\pi} \int_\Gamma \| (T^{(1)} R(\zeta))^n \| \, |d\zeta| \leqq \frac{1}{2\pi} \int_\Gamma \| T^{(1)} R(\zeta) \|^n \, |d\zeta| . \leqq \frac{1}{2\pi} \| T^{(1)} \|^n \int_\Gamma \| R(\zeta) \|^n \, |d\zeta| .$$

Substitution of (3.47) into (3.46) gives an estimate of $\hat{\lambda}^{(n)}$.

Remark 3.8. Under the last assumptions, the $T^{(1)}$ in (3.46–47) for $n \geqq 2$ may be replaced by $T^{(1)} - \alpha$ for any scalar α. This follows from the fact that the replacement of $T^{(1)}$ by $T^{(1)} - \alpha$ changes $T(\varkappa)$ only by the additive term $-\alpha\varkappa$ and does not affect $\hat{\lambda}^{(n)}$ for $n \geqq 2$. In particular, the $\| T^{(1)} \|$ in the last member of (3.47) may be replaced by

$$(3.48)\qquad a_0 = \min_\alpha \| T^{(1)} - \alpha \| .$$

5. The special case of a normal unperturbed operator

The foregoing results on the convergence radii and error estimates are much simplified in the special case in which X is a unitary space and T is normal. Then we have by I-(6.70)

$$(3.49)\qquad \| R(\zeta) \| = 1/\mathrm{dist}(\zeta, \Sigma(T))$$

for every $\zeta \in \mathrm{P}(T)$.

If we further assume that the $T^{(n)}$ satisfy the inequalities (3.7), then (3.8) gives

$$(3.50)\qquad r_0 = \min_{\zeta \in \Gamma} \left(\frac{a}{\mathrm{dist}(\zeta, \Sigma(T))} + c \right)^{-1}$$

as a lower bound for the convergence radii for $P(\varkappa)$ and $\hat{\lambda}(\varkappa)$. If we choose as Γ the circle $|\zeta - \lambda| = d/2$ where d is the isolation distance of the eigenvalue λ of T (see par. 2), we obtain

$$r_0 = \left(\frac{2a}{d} + c\right)^{-1}. \tag{3.51}$$

In the remainder of this paragraph we shall assume that $T^{(n)} = 0$ for $n \geqq 2$. Then we can take $c = 0$ and $a = \|T^{(1)}\|$, and (3.51) becomes

$$r_0 = d/2a = d/2\,\|T^{(1)}\|\,. \tag{3.52}$$

In other words, we have

Theorem 3.9.[1] *Let* X *be a unitary space, let* $T(\varkappa) = T + \varkappa T^{(1)}$ *and let* T *be normal. Then the power series for* $P(\varkappa)$ *and* $\hat{\lambda}(\varkappa)$ *are convergent if the "magnitude of the perturbation"* $\|\varkappa T^{(1)}\|$ *is smaller than half the isolation distance of the eigenvalue* λ *of* T.

Here the factor 1/2 is the best possible, as is seen from

Example 3.10. Consider Example 1.1, a) and introduce in X the unitary norm I-(6.6) with respect to the canonical basis for which $T(\varkappa)$ has the matrix a). It is then easy to verify that T is normal (even symmetric), $\|T^{(1)}\| = 1$ and $d = 2$ for each of the two eigenvalues ± 1 of T. But the convergence radii of the series (1.4) are exactly equal to $r_0 = 1$ given by (3.52).

Remark 3.11. $a = \|T^{(1)}\|$ in (3.52) can be replaced by a_0 given by (3.48) for the same reason as in Remark 3.8.

For the coefficients $\hat{\lambda}^{(n)}$, the formula (3.5) gives

$$|\hat{\lambda}^{(1)}| \leqq a\,, \quad |\hat{\lambda}^{(n)}| \leqq a_0^n \left(\frac{2}{d}\right)^{n-1}, \quad n \geqq 2\,, \tag{3.53}$$

for we have $\varrho = d/2$ for the Γ under consideration (see Remark 3.11). The formulas (3.46–47) lead to the same results (3.53) if the same Γ is used.

But (3.46) is able to give sharper estimates than (3.53) in some special cases. This happens, for example, when T is *symmetric* so that the eigenvalues of T are real. In this case, considering the eigenvalue λ of T, we can take as Γ the pair Γ_1, Γ_2 of straight lines perpendicular to the real axis passing through $(\lambda + \lambda_1)/2$ and $(\lambda + \lambda_2)/2$, where λ_1 and λ_2 denote respectively the largest eigenvalue of T below λ and the smallest one above. On setting

$$d_1 = \lambda - \lambda_1\,, \quad d_2 = \lambda_2 - \lambda\,, \tag{3.54}$$

we have

$$\|R(\zeta)\| = \left(\frac{d_j^2}{4} + \eta^2\right)^{-1/2}, \quad \zeta \in \Gamma_j\,, \quad j = 1, 2\,, \quad \eta = \operatorname{Im}\zeta\,. \tag{3.55}$$

[1] See T. Kato [1], Schäfke [4].

Hence (3.46–47) with Remark 3.8 give for $n \geqq 2$

(3.56)
$$|\hat{\lambda}^{(n)}| \leqq \frac{1}{2\pi} \min\left(1, \frac{N}{nm}\right) a_0^n \left[\int_{-\infty}^{\infty} \left(\frac{d_1^2}{4} + \eta^2\right)^{-\frac{n}{2}} d\eta + \int_{-\infty}^{\infty} \left(\frac{d_2^2}{4} + \eta^2\right)^{-\frac{n}{2}} d\eta\right]$$
$$= \min\left(1, \frac{N}{nm}\right) \frac{\Gamma\left(\frac{n-1}{2}\right)}{\sqrt{\pi}\,\Gamma\left(\frac{n}{2}\right)} a_0^n \cdot \frac{1}{2}\left[\left(\frac{2}{d_1}\right)^{n-1} + \left(\frac{2}{d_2}\right)^{n-1}\right],$$

in which Γ denotes the gamma function. It should be noted that if λ is the smallest or the largest eigenvalue of T, we can set $d_1 = \infty$ or $d_2 = \infty$, respectively, thereby improving the result.

It is interesting to observe that (3.56) is "the best possible", as is seen from the following example.

Example 3.12. Again take Example 3.10. We have $N = 2$, $\lambda = 1$, $m = 1$, $d_1 = 2$, $d_2 = \infty$ and $a_0 = \|T^{(1)}\| = 1$ and (3.56) gives

(3.57)
$$|\lambda^{(n)}| = |\hat{\lambda}^{(n)}| \leqq \frac{\Gamma\left(\frac{n-1}{2}\right)}{\sqrt{\pi}\, n\, \Gamma\left(\frac{n}{2}\right)}.$$

The correct eigenvalue $\lambda(\varkappa)$ is

(3.58)
$$\lambda(\varkappa) = (1 + \varkappa^2)^{1/2} = \sum_{p=0}^{\infty} \binom{1/2}{p} \varkappa^{2p}.$$

The coefficient $\lambda^{(n)}$ of $\varkappa^n$ in this series is $\binom{1/2}{n/2}$ for even n and is exactly equal to the right member of (3.57).

The factor $\alpha_n = \Gamma\left(\frac{n-1}{2}\right) \Big/ \sqrt{\pi}\,\Gamma\left(\frac{n}{2}\right)$ in (3.56) has the following values for smaller n:

n	α_n
2	$1 = 1.0000$
3	$2/\pi = 0.6366$
4	$1/2 = 0.5000$
5	$4/3\pi = 0.4244$
6	$3/8 = 0.3750$

α_n has the asymptotic value $n^{-1/2}$ for $n \to \infty$[1]. Thus (3.56) shows that $\hat{\lambda}^{(n)}$ is at most of the order

(3.59)
$$\text{const}\left(\frac{2}{d}\right)^{n-1} n^{-3/2}, \quad d = \min(d_1, d_2).$$

But $n^{-3/2}$ must be replaced by $n^{-1/2}$ if $N = \infty$, and this should be done for practical purposes even for finite N if it is large.

[1] The Γ-function has the asymptotic formula $\Gamma(x+1) = (2\pi)^{1/2} x^{x+1/2} e^{-x}(1 + O(x^{-1}))$.

Problem 3.13. (3.56) is sharper than (3.53).

Problem 3.14. Why do the equalities in (3.57) hold for even n in Example 3.10?

Let us now compare these results with the estimates given by the method of majorizing series. We first note that

$$D = 0\,, \quad \|P\| = 1\,, \quad \|S\| = 1/d\,, \tag{3.60}$$

since T is assumed to be normal. If we replace (3.15) by $p = \|T^{(1)}\| = a$, $q = \|T^{(1)}\|\,\|S\| = a/d$ and $s = \|S\| = 1/d$, (3.21) gives $r = d/4a$ as a lower bound for the convergence radii, a value just one-half of the value (3.52). On the other hand, the majorizing series (3.16) for $\hat{\lambda}(\varkappa) - \lambda$ becomes after these substitions

$$\begin{aligned} \Psi(\varkappa) &= a\varkappa + \frac{1}{d}\cdot\frac{2a^2\varkappa^2}{1 - \dfrac{2a\varkappa}{d} + \left(1 - \dfrac{4a\varkappa}{d}\right)^{1/2}} \\ &= \sum_{n=1}^{\infty} (-1)^{n-1}\binom{1/2}{n} 2^{2n-1}\frac{a^n}{d^{n-1}}\varkappa^n\,. \end{aligned} \tag{3.61}$$

The estimates for $\hat{\lambda}^{(n)}$ obtained from this majorizing series are

$$|\hat{\lambda}^{(1)}| \leqq a\,, \quad |\hat{\lambda}^{(2)}| \leqq a_0^2/d\,, \quad |\hat{\lambda}^{(3)}| \leqq 2a_0^3/d^2\,, \quad |\hat{\lambda}^{(4)}| \leqq 5a_0^4/d^3, \ldots \tag{3.62}$$

(replacement of a by a_0 is justified by Remark 3.8). (3.62) is sharper than (3.53) for $n \leqq 5$ but not for $n \geqq 6$. It is also sharper than (3.56) (with d_1 and d_2 replaced by d) for $n \leqq 3$ but not for $n \geqq 4$. In any case the majorizing series gives rather sharp estimates for the first several coefficients but not for later ones. In the same way the majorizing series (3.34) for the eigenfunction becomes (in the case $m = 1$)

$$\begin{aligned} \varphi(\varkappa) - \varphi &\ll \frac{2\varkappa\,\|T^{(1)}\varphi\|/d}{1 - \dfrac{2a\varkappa}{d} + \left(1 - \dfrac{4a\varkappa}{d}\right)^{1/2}} \\ &= \frac{\|T^{(1)}\varphi\|}{d}\sum_{n=1}^{\infty}(-1)^n\binom{1/2}{n+1}2^{2n+1}\left(\frac{a}{d}\right)^{n-1}\varkappa^n\,. \end{aligned} \tag{3.63}$$

For the first several coefficients of the expansion $\varphi(\varkappa) - \varphi = \sum \varkappa^n \varphi^{(n)}$, this gives (replacing a by a_0 as above)

$$\begin{aligned} &\|\varphi^{(1)}\| \leqq \|T^{(1)}\varphi\|/d\,, \quad \|\varphi^{(2)}\| \leqq 2\|T^{(1)}\varphi\|\,a_0/d^2\,, \\ &\|\varphi^{(3)}\| \leqq 5\|T^{(1)}\varphi\|\,a_0^2/d^3\,, \quad \|\varphi^{(4)}\| \leqq 14\|T^{(1)}\varphi\|\,a_0^3/d^4, \ldots\,. \end{aligned} \tag{3.64}$$

Here $\|T^{(1)}\varphi\|$ may also be replaced by $\min_\alpha \|(T^{(1)} - \alpha)\,\varphi\|$ for the same reason as above[1].

6. The enumerative method

An estimate of $\hat{\lambda}^{(n)}$ can also be obtained by *computing directly the number of terms* in the explicit formula (2.31)[2]. To illustrate the method, we assume for simplicity that X is a unitary space, $T^{(n)} = 0$ for $n \geqq 2$ and that T is normal. Recalling that $S^{(k)} = S^k$, $S^{(0)} = -P$ and $S^{(-k)} = D^k = 0$, $k > 0$, and noting (3.60), we obtain

$$|\hat{\lambda}^{(n)}| \leqq \frac{1}{nm}\sum_{k_1+\cdots+k_n=n-1} |\operatorname{tr} T^{(1)} S^{(k_1)} \ldots T^{(1)} S^{(k_n)}|\,. \tag{3.65}$$

---

[1] For related results see RELLICH [4] and SCHRÖDER [1]–[3].

[2] Cf. BLOCH [1].

Now the expression after the tr sign in (3.65) contains at least one factor $S^{(0)} = -P$, so that this expression is of rank $\leqq m$. Hence this trace can be majorized by m times the norm of this operator as we have done before, giving

$$|\hat{\lambda}^{(n)}| \leqq \frac{(2n-2)!}{n!\,(n-1)!} \frac{a_0^n}{d^{n-1}} = \beta_n\, a_0^n \left(\frac{2}{d}\right)^{n-1}, \quad n \geqq 2\,. \tag{3.66}$$

Here we have replaced $\|T^{(1)}\|$ by a_0 for the same reason as above. The numerical factor $2^{n-1}\, n\, \beta_n = (2n-2)!/((n-1)!)^2$ is the number of solutions of $k_1 + \cdots + k_n = n - 1$. For smaller values of n we have

n	β_n
2	$1 = 1.0000$
3	$1/2 = 0.5000$
4	$5/8 = 0.6250$
5	$7/8 = 0.8750$
6	$21/16 = 1.3125$

This shows that the estimate (3.66) is sharper than the simple estimate (3.53) only for $n \leqq 5$. In fact β_n has the asymptotic value $\pi^{-1/2}\, n^{-3/2}\, 2^{n-1}$ for $n \to \infty$, which is very large compared with unity.

Actually (3.66) is only a special case of the result obtained by the majorizing series: (3.66) is exactly the n-th coefficient of (3.61) for $n \geqq 2$ (again with a replaced by a_0).

Thus the enumerative method does not give any new result. Furthermore, it is more limited in scope than the method of majorizing series, for it is not easy to estimate effectively by enumeration the coefficients $\hat{\lambda}^{(n)}$ in more general cases.

Summing up, it may be concluded that the method of majorizing series gives in general rather sharp estimates in a closed form, especially for the first several terms of the series. In this method, however, it is difficult to take into account special properties (such as normality) of the operator. In such a special case the simpler method of contour integrals appears to be more effective. The estimates (3.50), (3.51) and (3.52) have so far been deduced only by this method.

§ 4. Similarity transformations of the eigenspaces and eigenvectors

1. Eigenvectors

In the preceding sections on the perturbation theory of eigenvalue problems, we have considered eigenprojections rather than eigenvectors (except in § 3.3) because the latter are not uniquely determined. In some cases, however, it is required to have an expression for eigenvectors $\varphi_h(\varkappa)$ of the perturbed operator $T(\varkappa)$ for the eigenvalue $\lambda_h(\varkappa)$. We shall deduce such formulas in the present section, but for the moment we shall be content with considering the *generalized eigenvectors*; by this we mean any non-zero vector belonging to the algebraic eigenspace $\mathsf{M}_h(\varkappa) = P_h(\varkappa)\mathsf{X}$ for the eigenvalue $\lambda_h(\varkappa)$ (see I-§ 5.4). Of course a generalized eigenvector is an eigenvector in the proper sense if $\lambda_h(\varkappa)$ is semisimple.

These eigenvectors can be obtained simply by setting

$$\varphi_{hk}(\varkappa) = P_h(\varkappa)\,\varphi_k\,, \tag{4.1}$$

where the φ_k are fixed, linearly independent vectors of X. For each h and k, $\varphi_{hk}(\varkappa)$ is an analytic function of $\varkappa$ representing a generalized eigenvector of $T(\varkappa)$ *as long as it does not vanish.* This way of constructing generalized eigenvectors, however, has the following inconveniences, apart from the fact that it is rather artificial. First, $\varphi_{hk}(\varkappa)$ may become zero for some $\varkappa$ which is not a singular point of $\lambda_h(\varkappa)$ or of $P_h(\varkappa)$. Second, the $\varphi_{hk}(\varkappa)$ for different k need not be linearly independent; in fact there do not exist more than m_h linearly independent eigenvectors for a $\lambda_h(\varkappa)$ with multiplicity m_h.

These inconveniences may be avoided to some extent by taking the vectors φ_k from the subspace $\mathsf{M}_h(\varkappa_0)$ where $\varkappa_0$ is a fixed, non-exceptional point. Thus we can take exactly m_h linearly independent vectors $\varphi_k \in \mathsf{M}_h(\varkappa_0)$, and the resulting m_h vectors (4.1) for fixed h are linearly independent for sufficiently small $|\varkappa - \varkappa_0|$ since $P_h(\varkappa)$ is holomorphic at $\varkappa = \varkappa_0$. Since $\dim \mathsf{M}_h(\varkappa) = m_h$, the m_h vectors $\varphi_{hk}(\varkappa)$ form a basis of $\mathsf{M}_h(\varkappa)$. In this way we have obtained a basis of $\mathsf{M}_h(\varkappa)$ depending holomorphically on $\varkappa$.

But this is still not satisfactory, since the $\varphi_{hk}(\varkappa)$ may not be linearly independent (and some of them may vanish) at some points $\varkappa$ which are not exceptional. We shall present two different methods to overcome these difficulties. In the next paragraph, we use a method based on differential equations to construct a global transformation function that maps $\mathsf{M}_h(0)$ onto $\mathsf{M}_h(\varkappa)$. Later (in par. 7) we shall give a more direct method for eliminating singularities of the $\varphi_{hk}(\varkappa)$ mentioned above.

2. Transformation functions[1]

Our problem can be set in the following general form. Suppose that a projection $P(\varkappa)$ in X is given, which is holomorphic in $\varkappa$ in a domain D of the $\varkappa$-plane. Then $\dim P(\varkappa)\,\mathsf{X} = m$ is constant by Lemma I-4.10. It is required to find m vectors $\varphi_k(\varkappa)$, $k = 1, \ldots, m$, which are holomorphic in $\varkappa$ and which form a basis of $\mathsf{M}(\varkappa) = P(\varkappa)\,\mathsf{X}$ for *all* $\varkappa \in$ D.

We may assume without loss of generality that $\varkappa = 0$ belongs to D. Our problem will be solved if we construct an operator-valued function $U(\varkappa)$ [hereafter called a *transformation function* for $P(\varkappa)$] with the following properties:

[1] The results of this and the following paragraphs were given by T. KATO [2] in connection with the adiabatic theorem in quantum mechanics.

(1) The inverse $U(\varkappa)^{-1}$ exists and both $U(\varkappa)$ and $U(\varkappa)^{-1}$ are holomorphic for $\varkappa \in \mathrm{D}$;

(2) $U(\varkappa)\, P(0)\, U(\varkappa)^{-1} = P(\varkappa)$.

The second property implies that $U(\varkappa)$ maps $\mathsf{M}(0)$ onto $\mathsf{M}(\varkappa)$ one to one (see I-§ 5.7). If $\{\varphi_k\}$, $k = 1, \ldots, m$, is a basis of $\mathsf{M}(0)$, it follows that the vectors

$$\varphi_k(\varkappa) = U(\varkappa)\, \varphi_k\,, \quad k = 1, \ldots, m\,, \tag{4.2}$$

form a basis of $\mathsf{M}(\varkappa)$, which solves our problem.

We shall now construct a $U(\varkappa)$ with the above properties under the assumption that D is *simply connected*. We have

$$P(\varkappa)^2 = P(\varkappa) \tag{4.3}$$

and by differentiation

$$P(\varkappa)\, P'(\varkappa) + P'(\varkappa)\, P(\varkappa) = P'(\varkappa)\,, \tag{4.4}$$

where we use $'$ to denote the differentiation $d/d\varkappa$. Multiplying (4.4) by $P(\varkappa)$ from the left and from the right and noting (4.3), we obtain [we write P in place of $P(\varkappa)$, etc., for simplicity]

$$P P' P = 0\,. \tag{4.5}$$

We now introduce the *commutator* Q of P' and P:

$$Q(\varkappa) = [P'(\varkappa), P(\varkappa)] = P'(\varkappa)\, P(\varkappa) - P(\varkappa)\, P'(\varkappa)\,. \tag{4.6}$$

Obviously P' and Q are holomorphic for $\varkappa \in \mathrm{D}$. It follows from (4.3), (4.5) and (4.6) that

$$P Q = -P P'\,, \quad Q P = P' P\,. \tag{4.7}$$

Hence (4.4) gives

$$P' = [Q, P]\,. \tag{4.8}$$

Let us now consider the differential equation

$$X' = Q(\varkappa)\, X \tag{4.9}$$

for the unknown $X = X(\varkappa)$. Since this is a *linear* differential equation, it has a unique solution holomorphic for $\varkappa \in \mathrm{D}$ when the initial value $X(0)$ is specified. This can be proved, for example, by the method of successive approximation in the same way as for a linear system of ordinary differential equations[1].

[1] In fact (4.9) is equivalent to a system of ordinary differential equations in a matrix representation. But it is more convenient to treat (4.9) as an operator differential equation without introducing matrices, in particular when $\dim \mathsf{X} = \infty$ (note that all the results of this paragraph apply to the infinite-dimensional case without modification). The standard successive approximation, starting from the zeroth approximation $X_0(\varkappa) = X(0)$, say, and proceeding by $X_n(\varkappa) = X(0) + \int_0^\varkappa Q(\varkappa)\, X_{n-1}(\varkappa)\, d\varkappa$, gives a sequence $X_n(\varkappa)$ of holomorphic operator-valued

Let $X(\varkappa) = U(\varkappa)$ be the solution of (4.9) for the initial condition $X(0) = 1$. The general solution of (4.9) can then be written in the form

$$X(\varkappa) = U(\varkappa)\, X(0)\,. \tag{4.10}$$

In fact (4.10) satisfies (4.9) and the initial condition; in view of the uniqueness of the solution it must be the required solution.

In quite the same way, the differential equation

$$Y' = -\,Y Q(\varkappa) \tag{4.11}$$

has a unique solution for a given initial value $Y(0)$. Let $Y = V(\varkappa)$ be the solution for $Y(0) = 1$. We shall show that $U(\varkappa)$ and $V(\varkappa)$ are inverse to each other. The differential equations satisfied by these functions give $(VU)' = V'U + VU' = -VQU + VQU = 0$. Hence VU is a constant and

$$V(\varkappa)\, U(\varkappa) = V(0)\, U(0) = 1\,. \tag{4.12}$$

This proves that $V = U^{-1}$ and, therefore, we have also

$$U(\varkappa)\, V(\varkappa) = 1\,. \tag{4.13}$$

We shall give an independent proof of (4.13), for later reference, for (4.13) is not implied by (4.12) if the underlying space is of infinite dimension. We have as above

$$(UV)' = QUV - UVQ = [Q, UV]\,. \tag{4.14}$$

This time it is not obvious that the right member of (4.14) is zero. But (4.14) is also a linear differential equation for $Z = UV$, and the uniqueness of its solution can be proved in the same way as for (4.9) and (4.11). Since $Z(\varkappa) = 1$ satisfies (4.14) as well as the initial condition $Z(0) = 1 = U(0)\, V(0)$, UV must coincide with Z. This proves (4.13)[1].

We now show that $U(\varkappa)$ satisfies the conditions (1), (2) required above. (1) follows from $U(\varkappa)^{-1} = V(\varkappa)$ implied by (4.12) and (4.13). To prove (2), we consider the function $P(\varkappa)\, U(\varkappa)$. We have

$$(PU)' = P'U + PU' = (P' + PQ)\, U = QPU \tag{4.15}$$

by (4.9) and (4.8). Thus $X = PU$ is a solution of (4.9) with the initial value $X(0) = P(0)$ and must coincide with $U(\varkappa)\, X(0) = U(\varkappa)\, P(0)$ by (4.10). This is equivalent to (2).

---

functions; it is essential here that D is simply-connected. It is easy to show that $X_n(\varkappa)$ converges to an $X(\varkappa)$ uniformly in each compact subset of D and that $X(\varkappa)$ is the unique holomorphic solution of (4.9) with the given initial value $X(0)$. Here it is essential that the operation $X \to Q(\varkappa)\, X$ is a *linear* operator acting in $\mathscr{B}(\mathsf{X})$.

[1] (4.13) can also be deduced from (4.12) by a more general argument based on the *stability of the index*; cf. X-§ 5.5.

Remark 4.1. In virtue of (4.7), the Q in the last member of (4.15) can be replaced by P'. Thus the function $W(\varkappa) = U(\varkappa)\, P(0) = P(\varkappa)\, U(\varkappa)$ satisfies the differential equation

$$W' = P'(\varkappa)\, W\,, \tag{4.16}$$

which is somewhat simpler than (4.9). Similarly, it is seen that $Z(\varkappa) = P(0)\; U(\varkappa)^{-1}$ satisfies

$$Z' = Z\, P'(\varkappa)\,. \tag{4.17}$$

Remark 4.2. $U(\varkappa)$ and $U(\varkappa)^{-1}$ can be continued analytically as long as this is possible for $P(\varkappa)$. But it may happen that they are not single-valued even when $P(\varkappa)$ is, if the domain of $\varkappa$ is not simply-connected. As an example, let $N = 2$ and

$$P(\varkappa) = \frac{1}{2}\begin{pmatrix} 1 & 1/\varkappa \\ \varkappa & 1 \end{pmatrix}, \quad \varkappa \in \mathrm{D} = \mathrm{C}\backslash\{0\}\,.$$

$P(\varkappa)$ is a projection: $P(\varkappa)^2 = P(\varkappa)$, holomorphic in $\varkappa \in \mathrm{D}$. A simple computation gives

$$Q(\varkappa) = [P'(\varkappa),\, P(\varkappa)] = \begin{pmatrix} -1/2\varkappa & 0 \\ 0 & 1/2\varkappa \end{pmatrix}.$$

Solving the differential equation (4.9) with the initial condition $U(1) = 1$ gives the transformation function

$$U(\varkappa) = \begin{pmatrix} \varkappa^{-1/2} & 0 \\ 0 & \varkappa^{1/2} \end{pmatrix}.$$

(Here we choose the initial point $\varkappa = 1$ since 0 is not in D.) Obviously $U(\varkappa)$ is analytic in $\varkappa \in \mathrm{D}$ but is not single-valued. One should note, however, that the single-valued function

$$U(\varkappa) = \begin{pmatrix} 1 & 0 \\ 0 & \varkappa \end{pmatrix} \quad \text{with} \quad U(\varkappa)^{-1} = \begin{pmatrix} 1 & 0 \\ 0 & 1/\varkappa \end{pmatrix}$$

is equally good as a transformation function. This shows that the recipe given above is not always a convenient method for constructing a transformation function. A different method to be given in par. 7 will lead to a single-valued $U(\varkappa)$.

Incidentally, $P(\varkappa)$ considered here is one of the two eigenprojections for the operator $T(\varkappa) = \begin{pmatrix} 0 & 1 \\ \varkappa^2 & 0 \end{pmatrix}$, which is obtained from Example 1.1, d) (see also Example 1.12, d)) by replacing $\varkappa$ with $\varkappa^2$.

Remark 4.3. The construction of $U(\varkappa)$ can be carried out even if $\varkappa$ is a real variable. In this case $P(\varkappa)$ need not be holomorphic; it suffices that $P'(\varkappa)$ exists and is continuous (or piecewise continuous). Then $U(\varkappa)$ has a continuous (piecewise continuous) derivative and satisfies (1), (2) except that it is not necessarily holomorphic in $\varkappa$.

Remark 4.4. The transformation function $U(\varkappa)$ is not unique for a given $P(\varkappa)$. Another $U(\varkappa)$ can be obtained from the result of I-§ 4.6, at least for sufficiently small $|\varkappa|$. Substituting $P(0)$, $P(\varkappa)$ for the P, Q of I-§ 4.6, we see from I-(4.38) that[1]

[1] (4.18) was given by Sz.-Nagy [1] in an apparently different form.

(4.18)
$$U(\varkappa) = [1 - (P(\varkappa) - P(0))^2]^{-1/2} [P(\varkappa) P(0) + (1 - P(\varkappa)) (1 - P(0))]$$

is a transformation function if $|\varkappa|$ is so small that $\|P(\varkappa) - P(0)\| < 1$. (4.18) is simpler than the $U(\varkappa)$ constructed above in that it is an algebraic expression in $P(\varkappa)$ and $P(0)$ while the other one was defined as the solution of a differential equation. But (4.18) has the inconvenience that it may not be defined for all $\varkappa \in D$.

3. Solution of the differential equation

Since we are primarily interested in the mapping of $\mathsf{M}(0)$ onto $\mathsf{M}(\varkappa)$ by the transformation function $U(\varkappa)$, it suffices to consider $W(\varkappa) = U(\varkappa) P(0)$ instead of $U(\varkappa)$. To determine W it suffices to solve the differential equation (4.16) for the initial condition $W(0) = P(0)$.

Let us solve this equation in the case where $P(\varkappa)$ is the total projection for the λ-group eigenvalues of $T(\varkappa)$. $P(\varkappa)$ has the form (2.3), so that

$$P'(\varkappa) = \sum_{n=0}^{\infty} (n+1) \varkappa^n P^{(n+1)} . \tag{4.19}$$

Since $W(0) = P(0) = P$, we can write

$$W(\varkappa) = P + \sum_{n=1}^{\infty} \varkappa^n W^{(n)} . \tag{4.20}$$

Substitution of (4.19) and (4.20) into (4.16) gives the following recurrence formulas for $W^{(n)}$:

$$n W^{(n)} = n P^{(n)} P + (n-1) P^{(n-1)} W^{(1)} + \cdots + P^{(1)} W^{(n-1)} , \quad n = 1, 2, \ldots . \tag{4.21}$$

The $W^{(n)}$ can be determined successively from (4.21) by making use of the expressions (2.12) for $P^{(n)}$. In this way we obtain

$$\begin{aligned} W^{(1)} &= P^{(1)} P , \\ W^{(2)} &= P^{(2)} P + \frac{1}{2} [P^{(1)}]^2 P , \\ W^{(3)} &= P^{(3)} P + \frac{2}{3} P^{(2)} P^{(1)} P + \frac{1}{3} P^{(1)} P^{(2)} P + \frac{1}{6} [P^{(1)}]^3 P . \end{aligned} \tag{4.22}$$

If the eigenvalue λ of T is semisimple, we have by (2.14)

$$\begin{aligned} W^{(1)} &= -S T^{(1)} P , \\ W^{(2)} &= -S T^{(2)} P + S T^{(1)} S T^{(1)} P - S^2 T^{(1)} P T^{(1)} P - \\ &\quad - \frac{1}{2} P T^{(1)} S^2 T^{(1)} P . \end{aligned} \tag{4.23}$$

In case the λ-group consists of a single eigenvalue (no splitting),

$\mathsf{M}(\varkappa) = P(\varkappa)\,\mathsf{X}$ is itself the algebraic eigenspace of $T(\varkappa)$ for this eigenvalue, and we have a set of generalized eigenvectors

$$\varphi_k(\varkappa) = W(\varkappa)\,\varphi_k\,, \quad k = 1, \ldots, m\,, \tag{4.24}$$

where $\{\varphi_k\}$, $k = 1, \ldots, m$, is a basis of $\mathsf{M}(0)$. According to the properties of $W(\varkappa)$, the m vectors (4.24) form a basis of $\mathsf{M}(\varkappa)$.

The function $Z(\varkappa) = P(0)\,U(\varkappa)^{-1}$ can be determined in the same way. Actually we need not solve the differential equation (4.17) independently. (4.17) differs from (4.16) only in the order of the multiplication of the unknown and the coefficient $P'(\varkappa)$. Thus the series $\sum \varkappa^n Z^{(n)}$ for $Z(\varkappa)$ is obtained from that of $W(\varkappa)$ by inverting the order of the factors in each term. This is true not only for the expression of $Z^{(n)}$ in terms of the $P^{(n')}$ but also in terms of P, S, $T^{(1)}$, $T^{(2)}$, ... as in (4.23). This is due to the fact that the expressions (2.12) for the $P^{(n)}$ are invariant under the inversion of the type described. This remark gives, under the same assumption that λ is semisimple,

$$\begin{aligned} Z^{(1)} &= -P\,T^{(1)}\,S\,, \\ Z^{(2)} &= -P\,T^{(2)}\,S + P\,T^{(1)}\,S\,T^{(1)}\,S - P\,T^{(1)}\,P\,T^{(1)}\,S^2 - \\ &\quad - \frac{1}{2}\,P\,T^{(1)}\,S^2\,T^{(1)}\,P\,. \end{aligned} \tag{4.25}$$

Remark 4.5. The other transformation function $U(\varkappa)$ given by (4.18) can also be written as a power series in $\varkappa$ under the same assumptions. As is easily seen, the expansion of $U(\varkappa)\,P$ coincides with that of $W(\varkappa)$ deduced above up to the order $\varkappa^2$ inclusive.

4. The transformation function and the reduction process

The transformation function $U(\varkappa)$ for the total projection $P(\varkappa)$ for the λ-group constructed above can be applied to the reduction process described in § 2.3. Since the λ-group eigenvalues are the eigenvalues of $T(\varkappa)$ in the invariant subspace $\mathsf{M}(\varkappa) = \mathsf{R}(P(\varkappa))$ and since $U(\varkappa)$ has the property $P(\varkappa) = U(\varkappa)\,P\,U(\varkappa)^{-1}$, the eigenvalue problem for $T(\varkappa)$ in $\mathsf{M}(\varkappa)$ is equivalent to the eigenvalue problem for the operator

$$U(\varkappa)^{-1}\,T(\varkappa)\,U(\varkappa) \tag{4.26}$$

considered in the subspace $\mathsf{M} = \mathsf{M}(0) = \mathsf{R}(P)$ (which is invariant under this operator). In fact, (4.26) has the same set of eigenvalues as $T(\varkappa)$, whereas the associated eigenprojections and eigennilpotents of (4.26) are related to those of $T(\varkappa)$ by the similarity transformation with $U(\varkappa)^{-1}$. Since we are interested in the λ-group only, it suffices to consider the operator

(4.27) $$P U(\varkappa)^{-1} T(\varkappa) U(\varkappa) P = Z(\varkappa) T(\varkappa) W(\varkappa)$$

with the Z and W introduced in the preceding paragraphs.

(4.27) is holomorphic in $\varkappa$ and in this sense is of the same form as the given operator $T(\varkappa)$. Thus the problem for the λ-group has been reduced to a problem within a fixed subspace M of X. This reduction of the original problem to a problem in a smaller subspace M has the advantage that M is independent of $\varkappa$, whereas in the reduction process considered in § 2.3 the subspace $\mathsf{M}(\varkappa)$ depends on $\varkappa$. For this reason the reduction to (4.27) is more complete at least theoretically, though it has the practical inconvenience that the construction of $U(\varkappa)$ is not simple.

In particular it follows that the weighted mean $\hat{\lambda}(\varkappa)$ of the λ-group eigenvalues is equal to m^{-1} times the trace of (4.27):

(4.28) $$\begin{aligned}\hat{\lambda}(\varkappa) &= m^{-1} \operatorname{tr} Z(\varkappa) T(\varkappa) W(\varkappa)\\ &= \lambda + m^{-1} \operatorname{tr} Z(\varkappa) (T(\varkappa) - \lambda) W(\varkappa) .\end{aligned}$$

Substitution of (4.23) and (4.25) for the coefficients of $W(\varkappa)$ and $Z(\varkappa)$ leads to the same results as in (2.33).

Problem 4.5a. Verify the last statement.

5. Simultaneous transformation for several projections

The $U(\varkappa)$ considered in par. 2 serves only for a single projection $P(\varkappa)$. We shall now consider several projections $P_h(\varkappa)$, $h = 1, \ldots, s$, satisfying the conditions

(4.29) $$P_h(\varkappa) P_k(\varkappa) = \delta_{hk} P_h(\varkappa) ,$$

and construct a transformation function $U(\varkappa)$ such that

(4.30) $$U(\varkappa) P_h(0) U(\varkappa)^{-1} = P_h(\varkappa) , \quad h = 1, \ldots, s .$$

As a consequence, we can find a basis $\{\varphi_{h1}(\varkappa), \ldots, \varphi_{hm_h}(\varkappa)\}$ of each subspace $\mathsf{M}_h(\varkappa) = \mathsf{R}(P_h(\varkappa))$ by setting

(4.31) $$\varphi_{hj}(\varkappa) = U(\varkappa) \varphi_{hj} ,$$

where $\{\varphi_{h1}, \ldots, \varphi_{hm_h}\}$ is a basis of $\mathsf{M}_h = \mathsf{M}_h(0)$.

As before we assume that the $P_h(\varkappa)$ are either holomorphic in a simply-connected domain D of the complex plane or continuously differentiable in an interval of the real line. We may assume that the set $\{P_h(\varkappa)\}$ is *complete* in the sense that

(4.32) $$\sum_{h=1}^{s} P_h(\varkappa) = 1 .$$

Otherwise we can introduce the projection $P_0(\varkappa) = 1 - \sum_{h=1}^{s} P_h(\varkappa)$; the new set $\{P_h(\varkappa)\}$, $h = 0, 1, \ldots, s$, will satisfy (4.29) as well as the completeness condition, and $U(\varkappa)$ satisfies (4.30) for the old set if and only if it does the same for the new set.

The construction of $U(\varkappa)$ is similar to that for a single $P(\varkappa)$. We define $U(\varkappa)$ as the solution of the differential equation (4.9) for the initial value $X(0) = 1$, in which $Q(\varkappa)$ is to be given now by

$$Q(\varkappa) = \frac{1}{2} \sum_{h=1}^{s} [P'_h(\varkappa), P_h(\varkappa)] = \sum_{h=1}^{s} P'_h(\varkappa)\, P_h(\varkappa) = - \sum_{h=1}^{s} P_h(\varkappa)\, P'_h(\varkappa)\,. \tag{4.33}$$

The equality of the three members of (4.33) follows from (4.32), which implies that $\Sigma P'_h P_h + \Sigma P_h P'_h = \Sigma (P_h^2)' = \Sigma P'_h = 0$. Note also that this $Q(\varkappa)$ coincides with (4.6) in the case of a single $P(\varkappa)$; the apparent difference due to the presence in (4.33) of the factor 1/2 arises from the fact that in (4.33) we have enlarged the single $P(\varkappa)$ to the pair $\{P(\varkappa), 1 - P(\varkappa)\}$ so as to satisfy the completeness condition (4.32).

The argument of par. 2 [in particular (4.15)] shows that (4.30) is proved if we can show that

$$P'_h(\varkappa) = [Q(\varkappa), P_h(\varkappa)]\,, \quad h = 1, \ldots, s\,. \tag{4.34}$$

To prove this we differentiate (4.29), obtaining

$$P'_h P_k + P_h P'_k = \delta_{hk} P'_h\,. \tag{4.35}$$

Multiplication from the left by P_h gives $P_h P'_h P_k + P_h P'_k = \delta_{hk} P_h P'_h = P_k P_h P'_h$, which may be written

$$-[P_h P'_h, P_k] = P_h P'_k\,. \tag{4.36}$$

Summation over $h = 1, \ldots, s$ gives (4.34) in virtue of (4.33).

Obviously a simultaneous transformation function $U(\varkappa)$ can be constructed in this way for the set of all eigenprojections $P_h(\varkappa)$ of $T(\varkappa)$ in any simply connected domain D of the $\varkappa$-plane in which the $P_h(\varkappa)$ are holomorphic[1].

6. Diagonalization of a holomorphic matrix function

Let $(\tau_{jk}(\varkappa))$ be an $N \times N$ matrix whose elements are holomorphic functions of a complex variable $\varkappa$. Under certain conditions such a matrix function can be diagonalized, that is, there is a matrix $(\gamma_{jk}(\varkappa))$ with elements holomorphic in $\varkappa$ such that

$$(\tau'_{jk}(\varkappa)) = (\gamma_{jk}(\varkappa))^{-1} (\tau_{jk}(\varkappa))\, (\gamma_{jk}(\varkappa)) \tag{4.37}$$

is a diagonal matrix for every $\varkappa$ considered.

[1] The transformation function $U(\varkappa)$ has important applications in the adiabatic theorem in quantum mechanics, for which we refer to T. Kato [2], Garrido [1], Garrido and Sancho [1].

This problem can be reduced to the one considered in this section. It suffices to regard the given matrix as an operator $T(\varkappa)$ acting in the space $\mathsf{X} = \mathbf{C}^N$ of numerical vectors and apply the foregoing results. If the value of $\varkappa$ is restricted to a simply-connected domain D containing no exceptional points, we can construct the $U(\varkappa)$ and therefore a basis (4.31) consisting of vector functions holomorphic for $\varkappa \in$ D and adapted to the set $\{P_h(\varkappa)\}$ of eigenprojections of $T(\varkappa)$. With respect to such a basis, the matrix representation of $T(\varkappa)$ takes the simple form described in I-§ 5.4. In particular it is a diagonal matrix with diagonal elements $\lambda_h(\varkappa)$ if the $D_h(\varkappa)$ are all identically zero [which happens, for example, if all $m_h = 1$ or $T(\varkappa)$ is normal for real $\varkappa$, say]. As is seen from I-§ 5.4, this is equivalent to the existence of a matrix function $(\gamma_{jk}(\varkappa))$ with the required property (4.37). Note that the column vectors $(\gamma_{1k}(\varkappa), \ldots, \gamma_{Nk}(\varkappa))$ are eigenvectors of the given matrix $(\tau_{jk}(\varkappa))$.

7. Geometric eigenspaces (eigenprojections)

So far we have been mainly concerned with the algebraic eigenprojections for a holomorphic family $T(\varkappa)$. We recall (see § 1.8) that in a simple subdomain D, the eigenvalues $\lambda_h(\varkappa)$ and the associated (algebraic) eigenprojections $P_h(\varkappa)$ are holomorphic in $\varkappa$. We now consider a geometric eigenprojection $Q_h(\varkappa)$, such that $Q_h(\varkappa)\,\mathsf{X}$ is the geometric eigenspace for $\lambda_h(\varkappa)$.

Theorem 4.6. *Each $\lambda_h(\varkappa)$ has a geometric eigenprojection $Q_h(\varkappa) \leqq P_h(\varkappa)$*[1] *holomorphic in $\varkappa \in$ D. Here it is understood that for certain exceptional values of $\varkappa$ belonging to an isolated subset of* D, *the true geometric eigenspace may be larger than $Q_h(\varkappa)\,\mathsf{X}$.*

Example 4.7. Let $T(\varkappa) = \begin{pmatrix} 0 & \varkappa \\ 0 & 0 \end{pmatrix}$ (see Examples 1.1, 1.12, c)). A geometric eigenprojection for the eigenvalue $\lambda(\varkappa) = 0$ is given by $Q(\varkappa) = \begin{pmatrix} 1 & a \\ 0 & 0 \end{pmatrix}$, where a is any complex number, while $P(\varkappa) = 1$ is the algebraic eigenprojection. The exceptional point mentioned in Theorem 4.6 is $\varkappa = 0$, where the true geometric eigenspace is the whole space X, which is larger than $Q(0)\mathsf{X}$. $Q(\varkappa)$ is not unique, since a is arbitrary.

Theorem 4.6 will be proved as an application of general theorems (to be proved in next paragraph) on holomorphic families of operators $T(\varkappa) \in \mathscr{B}(\mathsf{X}_1, \mathsf{X}_2)$, where the underlying domain of $\varkappa$ need not be simply-connected.

---

[1] For two projections P, Q, recall that $Q \leqq P$ stands for $Q = PQ = QP$ (see I-§ 3.4). Note that $Q \leqq P$ is equivalent to $Q\mathsf{X} \subset P\mathsf{X}$ and $Q^*\mathsf{X}^* \subset P^*\mathsf{X}^*$.

Theorem 4.8. (rank stability) *Let $T(\varkappa) \in \mathscr{B}(\mathsf{X}_1, \mathsf{X}_2)$ be a holomorphic family defined for $\varkappa \in \mathrm{D}$, where D is a (not necessarily simply-connected) domain in C. Then there is an integer $r \geqq 0$ (the maximum rank of $T(\varkappa)$), and an isolated subset* K *of* D, *such that* rank $T(\varkappa) = r$ *for* $\varkappa \notin \mathrm{K}$ *and* rank $T(\varkappa) < r$ *for $\varkappa \in \mathrm{K}$. There are r-dimensional projections $Q_j(\varkappa) \in \mathscr{B}(\mathsf{X}_j)$ $(j = 1, 2)$, holomorphic in $\varkappa \in \mathrm{D}$, such that $T(\varkappa) = T(\varkappa)\, Q_1(\varkappa) = Q_2(\varkappa)\, T(\varkappa)$ for all $\varkappa \in \mathrm{D}$. If $\varkappa \notin \mathrm{K}$, then $T(\varkappa)\, \mathsf{X}_1 = Q_2(\varkappa)\, \mathsf{X}_2$, $\mathsf{N}(T(\varkappa)) = (1 - Q_1(\varkappa))\, \mathsf{X}_1$, and $T(\varkappa)$ maps $Q_1(\varkappa)\, \mathsf{X}_1$ onto $Q_2(\varkappa)\, \mathsf{X}_2$ one to one.*

Theorem 4.9. *In* Theorem 4.8 *assume further that there are holomorphic families of projections $P_j(\varkappa) \in \mathscr{B}(\mathsf{X}_j)$ $(j = 1, 2)$ such that $T(\varkappa) = T(\varkappa)\, P_1(\varkappa) = P_2(\varkappa)\, T(\varkappa)$. Then the Q_j may be chosen in such a way that $Q_j(\varkappa) \leqq P_j(\varkappa)$ for $j = 1, 2$ and $\varkappa \in \mathrm{D}$.*

Proof of Theorem 4.6. Let $T_h(\varkappa) = (T(\varkappa) - \lambda_h(\varkappa))\, P_h(\varkappa)$. $T_h(\varkappa)$ is *essentially* the part of $T(\varkappa) - \lambda_h(\varkappa)$ in the reducing subspace $P_h(\varkappa)\, \mathsf{X}$ (in the sense explained in I-§ 5.1). According to the results of § 1.8, $T_h(\varkappa)$ is holomorphic in $\varkappa \in \mathrm{D}$ with $T_h(\varkappa) = T_h(\varkappa)\, P_h(\varkappa) = P_h(\varkappa)\, T_h(\varkappa)$. It follows from Theorems 4.8, 4.9 that rank $T_h(\varkappa) = r$ is constant for $\varkappa \notin \mathrm{K}$, where K is an isolated set, and that there are projections $Q_{hj}(\varkappa) \leqq P_h(\varkappa)$ $(j = 1, 2)$ such that $T_h(\varkappa)$ maps $Q_{h1}(\varkappa)\, \mathsf{X}$ onto $Q_{h2}(\varkappa)\, \mathsf{X}$ one to one for $\varkappa \notin \mathrm{K}$. If we recall that $T(\varkappa) - \lambda_h(\varkappa)$ maps $(1 - P_h(\varkappa))\, \mathsf{X}$ one to one onto itself, we see that the null space of $T(\varkappa) - \lambda_h(\varkappa)$ is exactly $(1 - Q_{h1}(\varkappa))\, P_h(\varkappa)\, \mathsf{X}$ except for $\varkappa \in \mathrm{K}$, for which the former is larger than the latter. Thus $(1 - Q_{h1}(\varkappa))\, P_h(\varkappa)$ is a geometric eigenprojection for $\lambda_h(\varkappa)$ required in Theorem 4.6.

Remark 4.10. Theorems 4.8, 4.9 are *global* in the sense that the $Q_j(\varkappa)$ are holomorphic in all of D. The global character becomes more apparent if we introduce transformation functions $U_j(\varkappa)$ associated with $Q_j(\varkappa)$, so that $Q_j(0) = U_j(\varkappa)^{-1}\, Q_j(\varkappa)\, U_j(\varkappa)$ (assuming, without loss of generality, that $0 \in \mathrm{D}$). Setting $T_0(\varkappa) = U_2(\varkappa)^{-1}\, T(\varkappa)\, U_1(\varkappa)$ gives

$$T_0(\varkappa) = T_0(\varkappa)\, Q_1(0) = Q_2(0)\, T_0(\varkappa). \tag{4.38}$$

This shows that $T(\varkappa)$ is "similar" to a family $T_0(\varkappa) \in \mathscr{B}(\mathsf{X}_1, \mathsf{X}_2)$ which has a fixed range $Q_2(0)\, \mathsf{X}_2$ and a fixed null space $(1 - Q_1(0))\, \mathsf{X}_1$, except for $\varkappa \in \mathrm{K}$.[1]

Construction of $T_0(\varkappa)$ given here requires transformation functions $U_j(\varkappa)$. Those given in par. 4 were available only when D is simply-connected. Another method for constructing the $U_j(\varkappa)$ without such a restriction will be given in par. 9.

---

[1] Similar results have been given by SILVERMAN and BUCY [1] in the case when $\varkappa$ is restricted to the real axis. It may be noted that for $\varkappa$ real (or close to the real axis) these results can be proved more easily by considering the selfadjoint families $T(\bar{\varkappa})^*\, T(\varkappa)$ and $T(\varkappa)\, T(\bar{\varkappa})^*$ (where the space X is assumed to be a unitary space without loss of generality); see § 5.7 below.

Remark 4.11. (relative inverse) In Theorem 4.8, assume that rank $T(\varkappa) = \text{const} = r$ for all $\varkappa \in \mathrm{D}$, so that K is empty. Since $T(\varkappa)$ maps $Q_1(\varkappa)\mathsf{X}_1$ onto $Q_2(\varkappa)\mathsf{X}_2$ one to one, the inverse operator $S(\varkappa)$ is defined on $Q_2(\varkappa)\mathsf{X}_2$ to $Q_1(\varkappa)\mathsf{X}_1$. This operator $S(\varkappa)$ is holomorphic in $\varkappa$ in the sense that it has holomorphic matrix elements with respect to bases in these subspaces that depend on $\varkappa$ holomorphically (because $T(\varkappa)$ has the same property). The existence of such bases is shown in the proof of Theorem 4.8 given below (see (4.42)). If we extend $S(\varkappa)$ to all of X_2 by setting $S(\varkappa)y = S(\varkappa)Q_2(\varkappa)y$, it is easy to see that $S(\varkappa) \in \mathscr{B}(\mathsf{X}_2, \mathsf{X}_1)$ is holomorphic in $\varkappa \in \mathrm{D}$ and satisfies $ST = Q_1$, $TS = Q_2$, $TST = T$ and $STS = S$. $S(\varkappa)$ is called a relative inverse[1] of $T(\varkappa)$.

8. Proof of Theorems 4.8, 4.9

For the proof of Theorems 4.8, 4.9, we need some elementary results of complex function theory.

Lemma 4.12[2]. *Let α_1, α_2 be two scalar-valued holomorphic functions in a domain $\mathrm{D} \subset \mathbf{C}$, with no common zeros. Then there exist two holomorphic functions β_1, β_2 in* D *such that $\alpha_1\beta_1 + \alpha_2\beta_2 = 1$ identically.*

Proof. Let Z_j be the zero set of α_j $(j = 1, 2)$. $\mathrm{Z}_1, \mathrm{Z}_2$ are disjoint by hypothesis. $1/\alpha_1\alpha_2$ is meromorphic in D, with poles in $\mathrm{Z}_1 \cup Z_2$. According to the Mittag-Leffler theorem[3], there is a meromorphic function γ_1, with poles in and only in Z_1, having the same principal part at each pole as $1/\alpha_1\alpha_2$ does. Then $\gamma_2 = 1/\alpha_1\alpha_2 - \gamma_1$ has poles only in Z_2, with the same principal parts as $1/\alpha_1\alpha_2$. Thus we have $\alpha_1\beta_1 + \alpha_2\beta_2 = 1$, where $\beta_1 = \alpha_2\gamma_2$ and $\beta_2 = \alpha_1\gamma_1$. But β_1 is holomorphic, since the poles of γ_2 are cancelled by the zeros of α_2, and similarly with β_2.

Lemma 4.13. *Let $u(\varkappa) \in \mathsf{X}$ be holomorphic in $\varkappa \in \mathrm{D}$ with no zeros. Then there is a function $f(\bar{\varkappa}) \in \mathsf{X}^*$ holomorphic in $\bar{\varkappa} \in \bar{\mathrm{D}}$ (the mirror image of* D *with respect to the real axis) such that $(u(\varkappa), f(\bar{\varkappa})) = 1$ identically for $\varkappa \in \mathrm{D}$.*

Proof. Fix $\varkappa_0 \in \mathrm{D}$. Since $u(\varkappa_0) \neq 0$, there is $\varphi \in \mathsf{X}^*$ such that $(u(\varkappa_0), \varphi) \neq 0$. Since $(u(\varkappa), \varphi)$ is holomorphic in $\varkappa$, its zeros form an isolated set $\mathrm{Z} \subset \mathrm{D}$. Since $u(\varkappa) \neq 0$ for each $\varkappa \in \mathrm{D}$, there is $\psi \in \mathsf{X}^*$ such that $(u(\varkappa), \psi) \neq 0$ for all $\varkappa \in \mathrm{Z}$. Indeed, X^* cannot be the *countable* union of the hyperplanes $\{u(\varkappa)\}^{\perp}$ for $\varkappa \in \mathrm{Z}$[4]. Thus the two scalar-valued holomorphic functions $(u(\varkappa), \varphi)$ and $(u(\varkappa), \psi)$ have no common zeros. It

---

[1] For relative inverses, see e.g. BART [1], GRAMSCH [2].

[2] Cf. HELMER [1], BURCKEL ⟦1⟧.

[3] See KNOPP ⟦2⟧, p. 37, where the theorem is given for the special case $\mathrm{D} = \mathbf{C}$. It is known to be valid for any domain D (see e.g. BURCKEL ⟦1⟧).

[4] If X is a general Banach space, one would need here Baire's category theorem (see Theorem III-1.14). If $\dim \mathsf{X} < \infty$, one may use an argument based on the Lebesgue measure.

follows from Lemma 4.12 that there are scalar-valued holomorphic functions α, β in D such that $\alpha(\varkappa)\,(u(\varkappa), \varphi) + \beta(\varkappa)\,(u(\varkappa), \psi) = 1$. Setting $f(\bar{\varkappa}) = \overline{\alpha(\varkappa)}\,\varphi + \overline{\beta(\varkappa)}\,\psi$ proves Lemma 4.13.

Lemma 4.14. *Let $v(\varkappa)$ be an* X*-valued meromorphic function in* D. *If v is not identically zero, there exists a scalar-valued meromorphic function α such that $u = \alpha\, v$ is holomorphic in* D *withous zeros.* (*The process of constructing u from v will be hereafter referred to as* regularization.)

Proof. It suffices that α has a zero [pole] of order k wherever v has a pole [zero] of order k. Such an α exists by a theorem of Weierstrass[1].

Proof of Theorem 4.8. The proof is a generalization of the usual Schmidt orthogonalization procedure (I-§ 6.3). We may assume, without loss of generality, that $0 \in \mathrm{D}$ and $\operatorname{rank} T(0) = r \geqq \operatorname{rank} T(\varkappa)$ for all $\varkappa \in \mathrm{D}$. Let $\varphi_1, \ldots \varphi_r \in \mathsf{X}_1$ such that $T(0)\,\varphi_j \in \mathsf{X}_2$, $j = 1, \ldots, r$, are linearly independent.

Set $v_1(\varkappa) = T(\varkappa)\,\varphi_1$. Since $v_1(\varkappa) \in \mathsf{X}_2$ is holomorphic for $\varkappa \in \mathrm{D}$ and not identically zero, we can regularize it to obtain a holomorphic function $u_1(\varkappa)$ without zeros (Lemma 4.14). By Lemma 4.13, there is an X_2^*-valued holomorphic function e_1 in $\bar{\mathrm{D}}$ such that

$$(u_1(\varkappa), e_1(\bar{\varkappa})) = 1 \quad \text{for} \quad \varkappa \in \mathrm{D} \tag{4.39}$$

Set $v_2(\varkappa) = T(\varkappa)\,\varphi_2$ and

$$w_2(\varkappa) = v_2(\varkappa) - (v_2(\varkappa), e_1(\bar{\varkappa}))\, u_1(\varkappa), \quad \varkappa \in \mathrm{D}. \tag{4.40}$$

w_2 is holomorphic in D and satisfies $(w_2(\varkappa), e_1(\bar{\varkappa})) = 0$ by (4.39). w_2 is not identically zero because $w_2(0) \neq 0$ by the linear independence of the $T(0)\,\varphi_j$. Hence we can regularize it to obtain a holomorphic function u_2 without zeros (Lemma 4.14). By Lemma 4.13, there is a holomorphic function $f_2(\bar{\varkappa}) \in \mathsf{X}_2^*$ in $\bar{\mathrm{D}}$ such that $(u_2(\varkappa), f_2(\bar{\varkappa})) = 1$. Then

$$e_2(\bar{\varkappa}) = f_2(\bar{\varkappa}) - \overline{(u_1(\varkappa), f_2(\bar{\varkappa}))}\, e_1(\bar{\varkappa}) \tag{4.41}$$

is holomorphic in $\bar{\varkappa} \in \bar{\mathrm{D}}$. Thus we have constructed u_j, e_k such that

$$(u_j(\varkappa), e_k(\bar{\varkappa})) = \delta_{jk}, \quad j, k = 1, \ldots, r, \quad \varkappa \in \mathrm{D}, \tag{4.42}$$

for $r = 2$.

This process can be continued to construct holomorphic functions $u_j(\varkappa) \in \mathsf{X}_2$ and $e_j(\bar{\varkappa}) \in \mathsf{X}_2^*$ on D and $\bar{\mathrm{D}}$, respectively, for $j = 1, 2, \ldots, r$, such that (4.42) is true for $j, k = 1, \ldots, r$.

In particular, the $u_j(\varkappa)$ are linearly independent for all $\varkappa \in \mathrm{D}$. Their span $\mathsf{M}(\varkappa)$ is identical with the span of the $v_j(\varkappa) = T(\varkappa)\,\varphi_j$, except for those isolated values of $\varkappa$ for which one of the $w_j(\varkappa)$ has a zero. Hence $\operatorname{rank} T(\varkappa) = r$ except for $\varkappa$ in an isolated set $\mathrm{K} \subset \mathrm{D}$.

---

[1] See Knopp ⟦2⟧, p. 6, where D = C is assumed. Again the theorem is true for any domain D (see e.g. Burckel ⟦1⟧).

On setting[1]

$$Q_2(\varkappa) = \sum_{j=1}^{r} (\ \ , e_j(\bar{\varkappa}))\, u_j(\varkappa) \in \mathscr{B}(\mathsf{X}_2), \quad \varkappa \in \mathrm{D}, \tag{4.43}$$

we obtain a holomorphic family of projections with global basis $u_j(\varkappa)$ such that $Q_2(\varkappa)\, \mathsf{X}_2 = \mathsf{M}(\varkappa)$.

If we apply the same construction for the holomorphic family $T(\bar{\varkappa})^*$ defined for $\varkappa \in \bar{\mathrm{D}}$, we arrive at a holomorphic family $Q_1(\varkappa)$ of projections in X_1 with global basis on D, such that $\mathsf{N}(T(\varkappa)) = (1 - Q_1(\varkappa))\, \mathsf{X}_1$ for $\varkappa \notin \mathrm{K}$ (see I-(3.37)). This completes the proof of Theorem 4.8.

Proof of Theorem 4.9. With the notation in the proof of Theorem 4.8, the assumption implies that $u_j(\varkappa) = P_2(\varkappa)\, u_j(\varkappa)$, since the same relation holds for $v_j(\varkappa) = T(\varkappa)\, \varphi_j$ and the $u_j(\varkappa)$ and the $v_j(\varkappa)$ have the same span. Consequently, the biorthogonality relation (4.42) does not change when the $e_k(\varkappa)$ are replaced with the $P_2(\varkappa)^*\, e_k(\varkappa)$. With this modification, we have $Q_2(\varkappa) \leqq P_2(\varkappa)$. Similarly we can prove the assertion regarding P_1, Q_1.

9. Remarks on projection families and transformation functions

When a holomorphic family $P(\varkappa) \in \mathscr{B}(\mathsf{X})$ of projections is given on a domain D, one may raise two questions. First, does there exist holomorphic vector functions $u_j(\varkappa) \in \mathsf{X}$, $j = 1, 2, \ldots, r = \dim P(\varkappa)$, which form a basis of $P(\varkappa)\mathsf{X}$ for each $\varkappa \in \mathrm{D}$ (global basis)?

If this is the case, it is easy to see that $P(\varkappa)$ can be expressed in the form

$$P(\varkappa) = \sum_{j=1}^{r} (\ \ , e_j(\bar{\varkappa}))\, u_j(\varkappa), \tag{4.44}$$

where

$$(u_j(\varkappa), e_k(\bar{\varkappa})) = \delta_{jk}, \quad e_k(\bar{\varkappa}) \in P(\varkappa)^* \mathsf{X}^*, \tag{4.45}$$

and the $e_k(\bar{\varkappa}) \in \mathsf{X}^*$ are holomorphic in $\bar{\varkappa} \in \bar{\mathrm{D}}$ and form a basis of $P(\varkappa)^*\mathsf{X}^*$. Indeed, the $e_k(\bar{\varkappa})$ are uniquely determined by (4.45) and easily seen to be holomorphic in $\bar{\varkappa}$. Conversely, an expression of the form (4.44) guarantees that $P(\varkappa)\mathsf{X}$ has a global basis $\{u_j(\varkappa)\}$. Note that this was the case for the $Q_j(\varkappa)$ in Theorems 4.8, 4.9.

Second, does there exist a holomorphic transformation function $U(\varkappa)$ satisfying

$$P(\varkappa) = U(\varkappa)\, P(0)\, U(\varkappa)^{-1} \tag{4.46}$$

assuming, for simplicity, that $0 \in \mathrm{D}$? Obviously, a positive answer to the second question implies the same for the first (see par. 4).

The two questions have positive answers if D is simply-connected, as was shown in par. 4. When D is not simply-connected, the first question still has a positive answer. This follows easily from the proof of Theorems 4.8, 4.9. Indeed Theorem 4.9, when applied to $T(\varkappa) = P(\varkappa) = P_1(\varkappa) = P_2(\varkappa)$, gives the result that $Q_1(\varkappa) = Q_2(\varkappa) = P(\varkappa)$. On the other hand, it was shown that $Q_2(\varkappa)$ has the form (4.44).

[1] $(\ \ , f)x$ is a linear operator of rank one which sends u into $(u, f)\,x$.

Incidentally, we note that these results are valid even when X is a Banach space, provided $\dim P(\varkappa) = r < \infty$.

The answer to the second question is also yes for any domain D if $\dim \mathsf{X} = N < \infty$. To see this, we have only to continue the generalized Schmidt procedure used in the proof of these theorems, with the set $v_j(\varkappa)$ $(j = 1, 2, \ldots, r)$ supplemented by $N - r$ functions $v_j(\varkappa) = \varphi_j$ $(j = r + 1, \ldots, N)$ such that the N vectors $v_j(0)$ are linearly independent. The map $U(\varkappa)$ that sends $u_j(0)$ to $u_j(\varkappa)$, $j = 1, \ldots, N$, is the required transformation function.

Finally, one may ask a more basic question. Given a (locally) holomorphic family $\mathsf{M}(\varkappa)$ of *subspaces* of X for $\varkappa \in$ D, does there exist a holomorphic family $P(\varkappa)$ of projections such that $P(\varkappa)\,\mathsf{X} = \mathsf{M}(\varkappa)$? Here $\mathsf{M}(\varkappa)$ ix *holomorphic* in $\varkappa$, by definition, if for each $\mu \in$ D there are a neighborhood D_μ of μ and holomoiphic functions $w_j(\varkappa) \in \mathsf{X}$ on D_μ that form a basis of $\mathsf{M}(\varkappa)$ for $\varkappa \in \mathrm{D}_\mu$.

Again the answer is yes, even for a Banach space X, provided $\dim \mathsf{M}(\varkappa) = r < \infty$[1]. We sketch a simple proof, which is a slight modification of the proof of Theorem 4.8. Fix r vectors $\psi_j \in \mathsf{X}^*$ $(j = 1, \ldots, r)$ such that $\det((w_j(0), \psi_k) \neq 0$, where $\{w_j(\varkappa)\}$ is a local basis of $\mathsf{M}(\varkappa)$ near $\varkappa = 0$. Then we can construct a new basis $\{v_i(\varkappa)\}$ of $\mathsf{M}(\varkappa)$, using linear combinations of the $w_j(\varkappa)$, such that

$$(v_j(\varkappa), \psi_k) = \delta_{jk}. \tag{4.47}$$

The $v_j(\varkappa)$ are thereby determined uniquely, and are meromorphic in $\varkappa$ in the given neighborhood of $\varkappa = 0$.

These functions $v_j(\varkappa)$ can be meromorphically continued to all $\varkappa \in$ D, since they are uniquely determined by (4.47). Thus the $v_j(\varkappa)$ form a *meromorphic global basis* of $\mathsf{M}(\varkappa)$. We can now apply the generalized Schmidt procedure used in the proof of Theorem 4.8 to the $v_j(\varkappa)$, to obtain the desired holomorphic global basis $u_j(\varkappa)$. Note that the regularization procedure given by Lemma 4.14 is applicable to a meromorphic function $v(\varkappa)$.

§ 5. Non-analytic perturbations

1. Continuity of the eigenvalues and the total projection

In the preceding sections we considered the eigenvalue problem for an operator $T(\varkappa) \in \mathscr{B}(\mathsf{X})$ holomorphic in $\varkappa$ and showed that its eigenvalues and eigenprojections are analytic functions of $\varkappa$. We now ask what conclusions can be drawn if we consider a more general type of dependence of $T(\varkappa)$ on $\varkappa$[2].

First we consider the case in which $T(\varkappa)$ is only assumed to be *continuous* in $\varkappa$. $\varkappa$ may vary in a domain D_0 of the complex plane or in an interval I of the real line. Even under this general assumption, some of the results of the foregoing sections remain essentially unchanged.

The resolvent $R(\zeta, \varkappa) = (T(\varkappa) - \zeta)^{-1}$ is now continuous in ζ and $\varkappa$ jointly in each domain for which ζ is different from any eigenvalue of $T(\varkappa)$. This is easily seen by modifying slightly the argument given in

[1] The existence of a global basis for $\mathsf{M}(\varkappa)$ was proved by SAPHAR [1].

[2] This question was discussed by RELLICH [1], [2], [8] (in greater detail in [8]) for symmetric operators.

§ 1.3; we need only to note that the $A(\varkappa) = T(\varkappa) - T$ in (1.11) is no longer holomorphic but tends to zero for $\varkappa \to 0$ $(T = T(0))$.

It follows that $R(\zeta, \varkappa)$ exists when ζ is in the resolvent set $\mathrm{P}(T)$ of T provided that $|\varkappa|$ is small enough to ensure that

$$\|T(\varkappa) - T\| < \|R(\zeta)\|^{-1} \quad (R(\zeta) = R(\zeta, 0))\,; \tag{5.1}$$

see (1.12). Furthermore, $R(\zeta, \varkappa) \to R(\zeta)$ for $\varkappa \to 0$ uniformly for ζ belonging to a compact subset of $\mathrm{P}(T)$.

Let λ be one of the eigenvalues of T, say with (algebraic) multiplicity m. Let Γ be a closed curve enclosing λ but no other eigenvalues of T. $\|R(\zeta)\|^{-1}$ has a positive minimum δ for $\zeta \in \Gamma$, and $R(\zeta, \varkappa)$ exists for all $\zeta \in \Gamma$ if $\|T(\varkappa) - T\| < \delta$. Consequently the operator $P(\varkappa)$ is again defined by (1.16) and is continuous in $\varkappa$ near $\varkappa = 0$. As in the analytic case, $P(\varkappa)$ is the total projection for the eigenvalues of $T(\varkappa)$ lying inside Γ. The continuity of $P(\varkappa)$ again implies that

$$\dim \mathsf{M}(\varkappa) = \dim \mathsf{M} = m\,, \quad \mathsf{M}(\varkappa) = P(\varkappa)\,\mathsf{X}\,, \quad \mathsf{M} = \mathsf{M}(0) = P\mathsf{X}\,, \tag{5.2}$$

where $P = P(0)$ is the eigenprojection of T for the eigenvalue λ. (5.2) implies that the sum of the multiplicities of the eigenvalues of $T(\varkappa)$ lying inside Γ is equal to m. These eigenvalues are again said to form the λ-group.

The same results are true for each eigenvalue λ_h of T. In any neighborhood of λ_h, there are eigenvalues of $T(\varkappa)$ with total multiplicity equal to the multiplicity m_h of λ_h provided that $|\varkappa|$ is sufficiently small. Since the sum of the m_h is N, there are no other eigenvalues of $T(\varkappa)$. This proves (and gives a precise meaning to) the proposition that *the eigenvalues of $T(\varkappa)$ are continuous in $\varkappa$.*

We assumed above that $T(\varkappa)$ is continuous in a domain of $\varkappa$. But the same argument shows that the eigenvalues of $T(\varkappa)$ and the total projection $P(\varkappa)$ are continuous at $\varkappa = 0$ if $T(\varkappa)$ is continuous at $\varkappa = 0$. To see this it suffices to notice that $R(\zeta, \varkappa) \to R(\zeta)$, $\varkappa \to 0$, uniformly for $\zeta \in \Gamma$. We may even replace $T(\varkappa)$ by a sequence $\{T_n\}$ such that $T_n \to T$, $n \to \infty$. Then it follows that the eigenvalues as well as the total projections of T_n tend to the corresponding ones of T for $n \to \infty$.

Summing up, we have

Theorem 5.1. *Let $T(\varkappa)$ be continuous at $\varkappa = 0$. Then the eigenvalues of $T(\varkappa)$ are continuous at $\varkappa = 0$. If λ is an eigenvalue of $T = T(0)$, the λ-group is well-defined for sufficiently small $|\varkappa|$ and the total projection $P(\varkappa)$ for the λ-group is continuous at $\varkappa = 0$. If $T(\varkappa)$ is continuous in a domain of the $\varkappa$-plane or in an interval of the real line, the resolvent $R(\zeta, \varkappa)$ is continuous in ζ and $\varkappa$ jointly in the sense stated above.*

2. The numbering of the eigenvalues

The fact proved above that the eigenvalues of $T(\varkappa)$ change continuously with $\varkappa$ when $T(\varkappa)$ is continuous in $\varkappa$ is not altogether simple since the number of eigenvalues of $T(\varkappa)$ is not necessarily constant. It is true that the same circumstance exists even in the analytic case, but there the number s of (different) eigenvalues is constant for non-exceptional $\varkappa$. In the general case now under consideration, this number may change with $\varkappa$ quite irregularly; the splitting and coalescence of eigenvalues may take place in a very complicated manner.

To avoid this inconvenience, it is usual to count the eigenvalues repeatedly according to their (algebraic) multiplicities as described in I-§ 5.4 (repeated eigenvalues). The repeated eigenvalues of an operator form an *unordered* N-tuple of complex numbers. Two such N-tuples $\mathfrak{S} = (\mu_1, \ldots, \mu_N)$ and $\mathfrak{S}' = (\mu_1', \ldots, \mu_N')$ may be considered close to each other if, *for suitable numbering of their elements*, the $|\mu_n - \mu_n'|$ are small for all $n = 1, \ldots, N$. We can even define the *distance* between such two N-tuples by

$$\text{dist}(\mathfrak{S}, \mathfrak{S}') = \min \max_n |\mu_n - \mu_n'| \tag{5.3}$$

where the min is taken over all possible renumberings of the elements of one of the N-tuples. For example, the distance between the triples (0, 0, 1) and (0, 1, 1) is equal to 1, though the set $\{0, 1\}$ of their elements is the same for the two triples. It is easy to verify that the distance thus defined satisfies the axioms of a distance function.

Problem 5.1a. $\mathfrak{S}(ST) = \mathfrak{S}(TS)$ for $T, S \in \mathscr{B}(\mathsf{X})$. [This is a refinement of $\Sigma(ST) = \Sigma(TS)$, see Problem I-5.4a. For the proof cf. Problem I-3.8a.]

The continuity of the eigenvalues of $T(\varkappa)$ given by Theorem 5.1 can now be expressed by saying that *the N-tuple $\mathfrak{S}(\varkappa)$ consisting of the repeated eigenvalues of $T(\varkappa)$ changes with $\varkappa$ continuously*. This means that the distance of $\mathfrak{S}(\varkappa)$ from $\mathfrak{S}(\varkappa_0)$ tends to zero for $\varkappa \to \varkappa_0$ for each fixed $\varkappa_0$.

The continuity thus formulated is the continuity of the repeated eigenvalues *as a whole*. It is a different question whether it is possible to define N single-valued, continuous functions $\mu_n(\varkappa)$, $n = 1, \ldots, N$, which for each $\varkappa$ represent the repeated eigenvalues of $T(\varkappa)$. Such a parametrization is in general impossible. This will be seen from Example 1.1, d), in which the two eigenvalues are $\pm\varkappa^{1/2}$; here it is impossible to define two single-valued continuous functions representing the two eigenvalues in a domain of the complex plane containing the branch point $\varkappa = 0$.

A parametrization is possible if either i) the parameter $\varkappa$ changes over an interval of the real line or ii) the eigenvalues are always real.

In case ii) it suffices to number the eigenvalues $\mu_n(\varkappa)$ in ascending (or descending) order:

$$\mu_1(\varkappa) \leqq \mu_2(\varkappa) \leqq \cdots \leqq \mu_N(\varkappa) . \tag{5.4}$$

It should be noted, however, that this way of numbering is not always convenient, for it can destroy the differentiability of the functions which may exist in a different arrangement.

The possibility of a parametrization in case i) is not altogether obvious. This is contained in the following theorem.

Theorem 5.2. *Let $\mathfrak{S}(\varkappa)$ be an unordered N-tuple of complex numbers, depending continuously on a real variable $\varkappa$ in a (closed or open) interval* I. *Then there exist N single-valued, continuous functions $\mu_n(\varkappa)$, $n = 1, \ldots, N$, the values of which constitute the N-tuple $\mathfrak{S}(\varkappa)$ for each $\varkappa \in \mathrm{I}$.* [*$(\mu_n(\varkappa))$ is called a representation of $\mathfrak{S}(\varkappa)$.*]

Proof. For convenience we shall say that a subinterval I_0 of I has *property* (A) if there exist N functions defined on I_0 with the properties stated in the theorem. What is required is to prove that I itself has property (A). We first show that, whenever two subintervals I_1, I_2 with property (A) have a common point, then their union $\mathrm{I}_0 = \mathrm{I}_1 \cup \mathrm{I}_2$ also has the same property. Let $(\mu_n^{(1)}(\varkappa))$ and $(\mu_n^{(2)}(\varkappa))$ be representations of $\mathfrak{S}(\varkappa)$ in I_1 and I_2, respectively, by continuous functions. We may assume that neither I_1 nor I_2 contains the other (otherwise the proof is trivial) and that I_1 lies to the left of I_2. For a fixed $\varkappa_0$ lying in the intersection of I_1 and I_2, we have $\mu_n^{(1)}(\varkappa_0) = \mu_n^{(2)}(\varkappa_0)$, $n = 1, \ldots, N$, after a suitable renumbering of $(\mu_n^{(2)})$, for both $(\mu_n^{(1)}(\varkappa_0))$ and $(\mu_n^{(2)}(\varkappa_0))$ represent the same $\mathfrak{S}(\varkappa_0)$. Then the functions $\mu_n^{(0)}(\varkappa)$ defined on I_0 by

$$\mu_n^{(0)}(\varkappa) = \begin{cases} \mu_n^{(1)}(\varkappa) , & \varkappa \leqq \varkappa_0 , \\ \mu_n^{(2)}(\varkappa) , & \varkappa \geqq \varkappa_0 , \end{cases} \tag{5.5}$$

are continuous and represent $\mathfrak{S}(\varkappa)$ on I_0.

It follows that, whenever a subinterval I′ has the property that each point of I′ has a neighborhood with property (A), then I′ itself has property (A).

With these preliminaries, we now prove Theorem 5.2 by induction. The theorem is obviously true for $N = 1$. Suppose that it has been proved for N replaced by smaller numbers and for any interval I. Let Γ be the set of all $\varkappa \in \mathrm{I}$ for which the N elements of $\mathfrak{S}(\varkappa)$ are identical, and let Δ be the complement of Γ in I. Γ is closed and Δ is open relative to I. Let us now show that each point of Δ has a neighborhood having property (A). Let $\varkappa_0 \in \Delta$. Since the N elements of $\mathfrak{S}(\varkappa_0)$ are not all identical, they can be divided into two separate groups with N_1 and N_2

elements, where $N_1 + N_2 = N$. In other words, $\mathfrak{S}(\varkappa_0)$ is composed of an N_1-tuple and an N_2-tuple with separate elements ("separate" means that there is no element of one group equal to an element of the other). The continuity of $\mathfrak{S}(\varkappa)$ implies that for sufficiently small $|\varkappa - \varkappa_0|$, $\mathfrak{S}(\varkappa)$ consists likewise of an N_1-tuple and an N_2-tuple each of which is continuous in $\varkappa$. According to the induction hypothesis, these N_1 and N_2-tuples can be represented in a neighborhood Δ' of $\varkappa_0$ by families of continuous functions $(\mu_1(\varkappa), \ldots, \mu_{N_1}(\varkappa))$ and $(\mu_{N_1+1}(\varkappa), \ldots, \mu_N(\varkappa))$, respectively. These N functions taken together then represent $\mathfrak{S}(\varkappa)$ in Δ'. In other words, Δ' has property (A).

Since Δ is open in I, it consists of at most countably many subintervals $\mathrm{I}_1, \mathrm{I}_2, \ldots$. Since each point of Δ has a neighborhood with property (A), it follows from the remark above that each component interval I_p has property (A). We denote by $\mu_n^{(p)}(\varkappa)$, $n = 1, \ldots, N$, the N functions representing $\mathfrak{S}(\varkappa)$ in I_p. For $\varkappa \in \Gamma$, on the other hand, $\mathfrak{S}(\varkappa)$ consists of N identical elements $\mu(\varkappa)$. We now define N functions $\mu_n(\varkappa)$, $n = 1, \ldots, N$, on I by

$$\mu_n(\varkappa) = \begin{cases} \mu_n^{(p)}(\varkappa), & \varkappa \in \mathrm{I}_p, \quad p = 1, 2, \ldots, \\ \mu(\varkappa), & \varkappa \in \Gamma. \end{cases} \tag{5.6}$$

These N functions represent $\mathfrak{S}(\varkappa)$ on the whole interval I. It is easy to verify that each $\mu_n(\varkappa)$ is continuous on I. This completes the induction and the theorem is proved.

3. Continuity of the eigenspaces and eigenvectors

Even when $T(\varkappa)$ is continuous in $\varkappa$, the eigenvectors or eigenspaces are not necessarily continuous. We have shown above that the total projection $P(\varkappa)$ for the λ-group is continuous, but $P(\varkappa)$ is defined only for small $|\varkappa|$ for which the λ-group eigenvalues are not too far from λ.

If $T(\varkappa)$ has N distinct eigenvalues $\lambda_h(\varkappa)$, $h = 1, \ldots, N$, for all $\varkappa$ in a simply connected domain of the complex plane or in an interval of the real line, we can define the associated eigenprojections $P_h(\varkappa)$ each of which is one-dimensional. Each $P_h(\varkappa)$ is continuous since it is identical with the total projection for the eigenvalue $\lambda_h(\varkappa)$. But $P_h(\varkappa)$ cannot in general be continued beyond a value of $\varkappa$ for which $\lambda_h(\varkappa)$ coincides with some other $\lambda_k(\varkappa)$. In this sense the eigenprojections behave more singularly than the eigenvalues. It should be recalled that even in the analytic case $P_h(\varkappa)$ may not exist at an exceptional point where $\lambda_h(\varkappa)$ is holomorphic [Example 1.12, f)]; but $P_h(\varkappa)$ has at most a pole in that case (see § 1.8). In the general case under consideration, the situation is much worse. That the eigenspaces can behave quite singularly even for a very smooth function $T(\varkappa)$ is seen from the following example due to RELLICH[1].

Example 5.3. Let $N = 2$ and

$$T(\varkappa) = e^{-\frac{1}{\varkappa^2}} \begin{pmatrix} \cos\frac{2}{\varkappa} & \sin\frac{2}{\varkappa} \\ \sin\frac{2}{\varkappa} & -\cos\frac{2}{\varkappa} \end{pmatrix}, \quad T(0) = 0\,. \tag{5.7}$$

$T(\varkappa)$ is not only continuous but is infinitely differentiable for all *real* values of $\varkappa$, and the same is true of the eigenvalues of $T(\varkappa)$, which are $\pm e^{-1/\varkappa^2}$ for $\varkappa \neq 0$ and zero for $\varkappa = 0$. But the associated eigenprojections are given for $\varkappa \neq 0$ by

$$\begin{pmatrix} \cos^2\frac{1}{\varkappa} & \cos\frac{1}{\varkappa}\sin\frac{1}{\varkappa} \\ \cos\frac{1}{\varkappa}\sin\frac{1}{\varkappa} & \sin^2\frac{1}{\varkappa} \end{pmatrix}, \begin{pmatrix} \sin^2\frac{1}{\varkappa} & -\cos\frac{1}{\varkappa}\sin\frac{1}{\varkappa} \\ -\cos\frac{1}{\varkappa}\sin\frac{1}{\varkappa} & \cos^2\frac{1}{\varkappa} \end{pmatrix}. \tag{5.8}$$

These matrix functions are continuous (even infinitely differentiable) in any interval not containing $\varkappa = 0$, but they cannot be continued to $\varkappa = 0$ as continuous functions. Furthermore, it is easily seen that there does not exist any eigenvector of $T(\varkappa)$ that is continuous in the neighborhood of $\varkappa = 0$ and that does not vanish at $\varkappa = 0$.

It should be remarked that (5.7) is a symmetric operator for each real $\varkappa$ (acting in C^2 considered a unitary space). In particular it is normal and, therefore, the eigenprojections would be *holomorphic* at $\varkappa = 0$ if $T(\varkappa)$ were *holomorphic* (Theorem 1.10). The example is interesting since it shows that this smoothness of the eigenprojections can be lost completely if the holomorphy of $T(\varkappa)$ is replaced by infinite differentiability.

4. Differentiability at a point

Let us now assume that $T(\varkappa)$ is not only continuous but differentiable. This does not in general imply that the eigenvalues of $T(\varkappa)$ are differentiable; in fact they need not be differentiable even if $T(\varkappa)$ is holomorphic [Example 1.1, d)]. However, we have

Theorem 5.4. *Let $T(\varkappa)$ be differentiable at $\varkappa = 0$. Then the total projection $P(\varkappa)$ for the λ-group is differentiable at $\varkappa = 0$:*

$$P(\varkappa) = P + \varkappa P^{(1)} + o(\varkappa)\,, \tag{5.9}$$

where[2] $P^{(1)} = -P T'(0) S - S T'(0) P$ *and S is the reduced resolvent of T for λ* (see I-§ 5.3). *If λ is a semisimple eigenvalue of T, the λ-group eigenvalues of $T(\varkappa)$ are differentiable at $\varkappa = 0$:*

$$\mu_j(\varkappa) = \lambda + \varkappa \mu_j^{(1)} + o(\varkappa)\,, \quad j = 1, \ldots, m\,, \tag{5.10}$$

where the $\mu_j(\varkappa)$ are the repeated eigenvalues of the λ-group and the $\mu_j^{(1)}$ are the repeated eigenvalues of $P T'(0) P$ in the subspace $\mathsf{M} = P\mathsf{X}$ (P is

[1] See RELLICH [1]; for convenience we modified the original example slightly.

[2] Here $o(\varkappa)$ denotes an operator-valued function $F(\varkappa)$ such that $\|F(\varkappa)\| = o(\varkappa)$ in the ordinary sense. See I-§ 4.7.

the eigenprojection of T for λ). If T is diagonable, all the eigenvalues are differentiable at $\varkappa = 0$.

Remark 5.5. The above theorem needs some comments. That the eigenvalues of $T(\varkappa)$ are differentiable means that the N-tuple $\mathfrak{S}(\varkappa)$ consisting of the repeated eigenvalues of $T(\varkappa)$ is differentiable, and similarly for the differentiability of the λ-group eigenvalues. That an (unordered) N-tuple $\mathfrak{S}(\varkappa)$ is differentiable at $\varkappa = 0$ means that $\mathfrak{S}(\varkappa)$ can be represented in a neighborhood of $\varkappa = 0$ by N functions $\mu_n(\varkappa)$, $n = 1, \ldots, N$, which are differentiable at $\varkappa = 0$. The N-tuple $\mathfrak{S}'(0)$ consisting of the $\mu_n'(0)$ is called the derivative of $\mathfrak{S}(\varkappa)$ at $\varkappa = 0$. It can be easily proved (by induction on N, say) that $\mathfrak{S}'(0)$ is thereby defined independently of the particular representation $(\mu_n(\varkappa))$ of $\mathfrak{S}(\varkappa)$. If $\mathfrak{S}(\varkappa)$ is differentiable at each $\varkappa$ and $\mathfrak{S}'(\varkappa)$ is continuous, $\mathfrak{S}(\varkappa)$ is said to be continuously differentiable.

Note that $\mathfrak{S}(0)$ and $\mathfrak{S}'(0)$ together need not determine the behavior of $\mathfrak{S}(\varkappa)$ even in the immediate neighborhood of $\varkappa = 0$. For example, $\mathfrak{S}_1(\varkappa) = (\varkappa, 1 - \varkappa)$ and $\mathfrak{S}_2(\varkappa) = (-\varkappa, 1 + \varkappa)$ have the common value $(0, 1)$ and the common derivative $(1, -1)$ at $\varkappa = 0$.

Proof of Theorem 5.4. We first note that the resolvent $R(\zeta, \varkappa)$ is differentiable at $\varkappa = 0$, for it follows from I-(4.28) that

$$\left[\frac{\partial}{\partial \varkappa} R(\zeta, \varkappa)\right]_{\varkappa=0} = -R(\zeta)\, T'(0)\, R(\zeta)\,. \tag{5.11}$$

Here it should be noted that the derivative (5.11) exists *uniformly* for ζ belonging to a compact subset of $\mathrm{P}(T)$, for $R(\zeta, \varkappa) \to R(\zeta)$, $\varkappa \to 0$, holds uniformly (see par. 1). Considering the expression (1.16) for $P(\varkappa)$, it follows that $P(\varkappa)$ is differentiable at $\varkappa = 0$, with

$$\begin{aligned} P'(0) &= -\frac{1}{2\pi i} \int_\Gamma \left[\frac{\partial}{\partial \varkappa} R(\zeta, \varkappa)\right]_{\varkappa=0} d\zeta = \frac{1}{2\pi i} \int_\Gamma R(\zeta)\, T'(0)\, R(\zeta)\, d\zeta \\ &= -P\,T'(0)\,S - S\,T'(0)\,P = P^{(1)} \quad [\text{cf. } (2.14)]\,. \end{aligned} \tag{5.12}$$

This proves (5.9).

As in the analytic case, if λ is semisimple the λ-group eigenvalues of $T(\varkappa)$ are of the form

$$\mu_j(\varkappa) = \lambda + \varkappa\, \mu_j^{(1)}(\varkappa)\,, \qquad j = 1, \ldots, m\,, \tag{5.13}$$

where the $\mu_j^{(1)}(\varkappa)$ are the (repeated) eigenvalues of the operator

$$\tilde{T}^{(1)}(\varkappa) = \varkappa^{-1}(T(\varkappa) - \lambda)\, P(\varkappa) = -\frac{\varkappa^{-1}}{2\pi i} \int_\Gamma (\zeta - \lambda)\, R(\zeta, \varkappa)\, d\zeta\,, \tag{5.14}$$

in the subspace $\mathsf{M}(\varkappa) = P(\varkappa)\,\mathsf{X}$ [see (2.37)]. Since $T(\varkappa)$ and $P(\varkappa)$ are differentiable at $\varkappa = 0$ and $(T - \lambda)\,P = 0$ if λ is semisimple, $\tilde{T}^{(1)}(\varkappa)$ is continuous at $\varkappa = 0$ if we set

$$\tilde{T}^{(1)}(0) = T'(0)\, P + (T - \lambda)\, P'(0) = P\, T'(0)\, P \tag{5.15}$$

where we have used (5.12) and $(T - \lambda)\, S = 1 - P$ [see (2.11)]. Hence the eigenvalues of $\tilde{T}^{(1)}(\varkappa)$ are continuous at $\varkappa = 0$. In particular, the eigenvalues $\mu_j^{(1)}(\varkappa)$ of $\tilde{T}^{(1)}(\varkappa)$ in the invariant subspace $\mathsf{M}(\varkappa)$ are continuous at $\varkappa = 0$ (see Theorem 5.1), though they need not be continuous for $\varkappa \neq 0$. In view of (5.13), this proves (5.10).

5. Differentiability in an interval

So far we have been considering the differentiability of the eigenvalues and eigenprojections at a single point $\varkappa = 0$. Let us now consider the differentiability in a domain of $\varkappa$, assuming that $T(\varkappa)$ is differentiable in this domain. If this is a domain of the complex plane, $T(\varkappa)$ is necessarily holomorphic; since this case has been discussed in detail, we shall henceforth assume that $T(\varkappa)$ is defined and differentiable in an interval I of the *real line*[1].

According to Theorem 5.4, the N-tuple $\mathfrak{S}(\varkappa)$ of the repeated eigenvalues of $T(\varkappa)$ is differentiable for $\varkappa \in$ I *provided that $T(\varkappa)$ is diagonable for $\varkappa \in$ I*. However, it is by no means obvious that there exist N functions $\mu_n(\varkappa)$, $n = 1, \ldots, N$, representing the repeated eigenvalues of $T(\varkappa)$, which are single-valued and differentiable on the whole of I. Actually this is true, as is seen from the following general theorem.

Theorem 5.6. *Let $\mathfrak{S}(\varkappa)$ be an unordered N-tuple of complex numbers depending on a real variable $\varkappa$ in an interval* I, *and let $\mathfrak{S}(\varkappa)$ be differentiable at each $\varkappa \in$ I (in the sense of* Remark 5.5*). Then there exist N complex-valued functions $\mu_n(\varkappa)$, $n = 1, \ldots, N$, representing the N-tuple $\mathfrak{S}(\varkappa)$ for $\varkappa \in$ I, each of which is differentiable for $\varkappa \in$ I.*

Proof. A subinterval of I for which there exist N functions of the kind described will be said to have property (B); it is required to prove that I itself has property (B). It can be shown, as in the proof of Theorem 5.2, that for any overlapping[2] subintervals I_1, I_2 with property (B), their union $\mathrm{I}_1 \cup \mathrm{I}_2$ has also property (B). The only point to be noted is that care must be taken in the renumbering of the $\mu_n^{(2)}(\varkappa)$ to make sure that the continuation (5.5) preserves the differentiability of the functions at $\varkappa = \varkappa_0$; this is possible owing to the assumption that $\mathfrak{S}(\varkappa)$ is differentiable at $\varkappa = \varkappa_0$.

---

[1] The differentiability of eigenvalues was investigated in detail in RELLICH [8] in the case when $T(\varkappa)$ is symmetric for each real $\varkappa$. It should be noted that the problem is far from trivial even in that special case.

[2] Here "overlapping" means that the two intervals have common interior points.

With this observation, the proof is carried out in the same way as in Theorem 5.2. A slight modification is necessary at the final stage, for the functions defined by (5.6) may have discontinuities of derivatives at an *isolated point* of Γ. To avoid this, we proceed as follows. An isolated point $\varkappa_0$ of Γ is either a boundary point of I or a common boundary of an I_p and an I_q. In the first case we have nothing to do. In the second case, it is easy to "connect" the two families $(\mu_n^{(p)}(\varkappa))$ and $(\mu_n^{(q)}(\varkappa))$ "smoothly" by a suitable renumbering of the latter, for these two families of differentiable functions represent $\mathfrak{S}(\varkappa)$ at the different sides of $\varkappa = \varkappa_0$ and $\mathfrak{S}(\varkappa)$ is differentiable at $\varkappa = \varkappa_0$. It follows that the interval consisting of I_p, I_q and $\varkappa_0$ has property (B).

Let Γ' be the set of isolated points of Γ. $\Delta \cup \Gamma'$ is relatively open in I and consists of (at most) countably many subintervals I'_k. Each I'_k consists in turn of countably many subintervals of the form I_p joined with one another at a point of Γ' in the manner stated above. By repeated applications of the connection process just described, the functions representing $\mathfrak{S}(\varkappa)$ in these I_p can be connected to form a family of N functions differentiable on I'_k. This shows that each I'_k has property (B).

The construction of N differentiable functions $\mu_n(\varkappa)$ representing $\mathfrak{S}(\varkappa)$ on the whole interval I can now be carried out by the method (5.6), in which the I_p and Γ should be replaced by the I'_k and Γ'' respectively, Γ'' being the complement of Γ' in Γ. The differentiability at a point $\varkappa_0$ of Γ'' of the $\mu_n(\varkappa)$ thus defined follows simply from the fact that the derivative $\mathfrak{S}'(\varkappa_0)$ consists of N identical elements, just as does $\mathfrak{S}(\varkappa_0)$. This completes the proof of Theorem 5.6.

Theorem 5.7. *If in* Theorem 5.6 *the derivative* $\mathfrak{S}'(\varkappa)$ *is continuous, the* N *functions* $\mu_n(\varkappa)$ *are continuously differentiable on* I.

Proof. Suppose that the real part of $\mu'_n(\varkappa)$ is discontinuous at $\varkappa = \varkappa_0$. According to a well-known result in differential calculus, the values taken by $\operatorname{Re}\mu'_n(\varkappa)$ in any neighborhood of $\varkappa_0$ cover an interval of length larger than a fixed positive number[1]. But this is impossible if $\mathfrak{S}'(\varkappa)$ is continuous, for any value of $\mu'_n(\varkappa)$ is among the N elements of $\mathfrak{S}'(\varkappa)$. For the same reason $\operatorname{Im}\mu'_n(\varkappa)$ cannot be discontinuous at any point.

Remark 5.8. We have seen (Theorem 5.4) that the eigenvalues of $T(\varkappa)$ are differentiable on I if $T(\varkappa)$ is differentiable and diagonable for $\varkappa \in$ I.

---

[1] Let a real-valued function $f(t)$ of a real variable t be differentiable for $a \leq t \leq b$; then $f'(t)$ takes in this interval all values between $\alpha = f'(a)$ and $\beta = f'(b)$. To prove this, we may assume $\alpha < \beta$. For any $\gamma \in (\alpha, \beta)$ set $g(t) = f(t) - \gamma t$. Then $g'(a) = \alpha - \gamma < 0$, $g'(b) = \beta - \gamma > 0$, so that the continuous function $g(t)$ takes a minimum at a point $t = c \in (a, b)$. Hence $g'(c) = 0$ or $f'(c) = \gamma$. If $f'(t)$ is discontinuous at $t = t_0$, then there is $\varepsilon > 0$ such that, in any neighborhood of t_0, there exist t_1, t_2 with $|f'(t_1) - f'(t_2)| > \varepsilon$. It follows from the above result that $f'(t)$ takes all values between $f'(t_1)$ and $f'(t_2)$. Hence the values of $f'(t)$ in any neighborhood of t_0 covers an interval of length larger than ε.

It is quite natural, then, to conjecture that the eigenvalues are continuously differentiable if $T(\varkappa)$ is continuously differentiable and diagonable. *But this is not true*, as is seen from the following example[1].

Example 5.9. Let $N = 2$ and

$$(5.16)\qquad T(\varkappa) = \begin{pmatrix} |\varkappa|^\alpha & |\varkappa|^\alpha - |\varkappa|^\beta \left(2 + \sin \frac{1}{|\varkappa|}\right) \\ -|\varkappa|^\alpha & -|\varkappa|^\alpha \end{pmatrix}, \quad \varkappa \neq 0; \quad T(0) = 0 .$$

$T(\varkappa)$ is continuously differentiable for all real $\varkappa$ if $\alpha > 1$ and $\beta > 2$. The two eigenvalues of $T(\varkappa)$ are

$$(5.17)\qquad \mu_\pm(\varkappa) = \pm |\varkappa|^{\frac{\alpha+\beta}{2}} \left(2 + \sin \frac{1}{|\varkappa|}\right)^{\frac{1}{2}}, \quad \varkappa \neq 0, \quad \mu_\pm(0) = 0 .$$

Since the $\mu_\pm(\varkappa)$ are different from each other for $\varkappa \neq 0$, $T(\varkappa)$ is diagonable, and $T(0) = 0$ is obviously diagonable. The $\mu_\pm(\varkappa)$ are differentiable everywhere, as the general theory requires. But their derivatives are discontinuous at $\varkappa = 0$ if $\alpha + \beta \leqq 4$. This simple example shows again that a non-analytic perturbation can be rather pathological in behavior.

Remark 5.10. If $T(\varkappa)$ is continuously differentiable in a neighborhood of $\varkappa = 0$ and λ is a semisimple eigenvalue of $T(0)$, the total projection $P(\varkappa)$ for the λ-group is continuously differentiable and $\tilde{T}^{(1)}(\varkappa)$ is continuous in a neighborhood of $\varkappa = 0$, as is seen from (5.12) and (5.14).

6. Asymptotic expansion of the eigenvalues and eigenvectors

The differentiability of the eigenvalues considered in the preceding paragraphs can be studied from a somewhat different point of view. Theorem 5.4 may be regarded as giving an *asymptotic* expansion of the eigenvalues $\mu_j(\varkappa)$ up to the first order of $\varkappa$ when $T(\varkappa)$ has the asymptotic form $T(\varkappa) = T + \varkappa T' + o(\varkappa)$, where $T' = T'(0)$. Going to the second order, we can similarly inquire into the asymptotic behavior of the eigenvalues when $T(\varkappa)$ has the asymptotic form $T + \varkappa T^{(1)} + \varkappa^2 T^{(2)} + o(\varkappa^2)$. In this direction, the extension of Theorem 5.4 is rather straightforward.

Theorem 5.11. *Let* $T(\varkappa) = T + \varkappa T^{(1)} + \varkappa^2 T^{(2)} + o(\varkappa^2)$ *for* $\varkappa \to 0$. *Let* λ *be an eigenvalue of* T *with the eigenprojection* P. *Then the total projection* $P(\varkappa)$ *for the* λ*-group eigenvalues of* $T(\varkappa)$ *has the form*

$$(5.18)\qquad P(\varkappa) = P + \varkappa P^{(1)} + \varkappa^2 P^{(2)} + o(\varkappa^2)$$

where P, $P^{(1)}$ *and* $P^{(2)}$ *are given by* (2.13). *If the eigenvalue* λ *of* T *is semisimple and if* $\lambda_j^{(1)}$ *is an eigenvalue of* $P T^{(1)} P$ *in* $P\mathsf{X}$ *with the eigen-*

[1] The operator of this example is not symmetric. For symmetric operators a slightly better behavior is expected; see Theorem 6.8 and Remark 6.9a.

projection $P_j^{(1)}$, *then* $T(\varkappa)$ *has exactly* $m_j^{(1)} = \dim P_j^{(1)}$ *repeated eigenvalues (the* $\lambda + \varkappa\,\lambda_j^{(1)}$*-group) of the form* $\lambda + \varkappa\,\lambda_j^{(1)} + o(\varkappa)$. *The total eigenprojection* $P_j^{(1)}(\varkappa)$ *for this group has the form*

$$P_j^{(1)}(\varkappa) = P_j^{(1)} + \varkappa\,P_j^{(11)} + o(\varkappa)\,. \tag{5.19}$$

If, in addition, the eigenvalue $\lambda_j^{(1)}$ *of* $P\,T^{(1)}\,P$ *is semisimple, then* $P_j^{(11)}$ *is given by* (2.47) *and the* $m_j^{(1)}$ *repeated eigenvalues of the* $\lambda + \varkappa\,\lambda_j^{(1)}$*-group have the form*

$$\mu_{jk}(\varkappa) = \lambda + \varkappa\,\lambda_j^{(1)} + \varkappa^2\,\mu_{jk}^{(2)} + o(\varkappa^2)\,, \quad k = 1, \ldots, m_j^{(1)}\,, \tag{5.20}$$

where $\mu_{jk}^{(2)}$, $k = 1, \ldots, m_j^{(1)}$, *are the repeated eigenvalues of* $P_j^{(1)}\,\tilde{T}^{(2)}\,P_j^{(1} = P_j^{(1)}\,T^{(2)}\,P_j^{(1)} - P_j^{(1)}\,T^{(1)}\,S\,T^{(1)}\,P_j^{(1)}$ *in the subspace* $P_j^{(1)}\mathsf{X}$.

Proof. The possibility of the asymptotic expansion of $T(\varkappa)$ up to the second order implies the same for the resolvent:

$$\begin{aligned} R(\zeta, \varkappa) &= R(\zeta) - R(\zeta)\,(T(\varkappa) - T)\,R(\zeta) + R(\zeta)\,(T(\varkappa) - T)\,\times \\ &\qquad \times R(\zeta)\,(T(\varkappa) - T)\,R(\zeta) + \cdots + \\ &= R(\zeta) - \varkappa\,R(\zeta)\,T^{(1)}\,R(\zeta) + \varkappa^2[-R(\zeta)\,T^{(2)}\,R(\zeta) + \\ &\qquad + R(\zeta)\,T^{(1)}\,R(\zeta)\,T^{(1)}\,R(\zeta)] + o(\varkappa^2)\,. \end{aligned} \tag{5.21}$$

Here $o(\varkappa^2)$ is uniform in ζ in each compact subset of $\mathrm{P}(T)$. Substitution of (5.21) into (1.16) yields (5.18), just as in the analytic case.

It follows that if λ is semisimple, the $\tilde{T}^{(1)}(\varkappa)$ of (5.14) has the form

$$\tilde{T}^{(1)}(\varkappa) = P\,T^{(1)}\,P + \varkappa\,\tilde{T}^{(2)} + o(\varkappa) \tag{5.22}$$

where $\tilde{T}^{(2)}$ is given by (2.20). The application of Theorem 5.4 to $\tilde{T}^{(1)}(\varkappa)$ then leads to the result of the theorem. Again the calculation of the terms up to the order $\varkappa^2$ is the same as in the analytic case.

7. Operators depending on several parameters

So far we have been concerned with an operator $T(\varkappa)$ depending on a single parameter $\varkappa$. Let us now consider an operator $T(\varkappa_1, \varkappa_2)$ depending on two variables $\varkappa_1$, $\varkappa_2$ which may be complex or real.

There is nothing new in regard to the continuity of the eigenvalues. The eigenvalues are continuous in $\varkappa_1$, $\varkappa_2$ (in the sense explained in par. 1,2) if $T(\varkappa_1, \varkappa_2)$ is continuous. Again, the same is true for partial differentiability (when the variables $\varkappa_1$, $\varkappa_2$ are real). But something singular appears concerning *total differentiability*. The total differentiability of $T(\varkappa_1, \varkappa_2)$ does not necessarily imply the same for the eigenvalues even if $T(\varkappa_1, \varkappa_2)$ is diagonable (cf. Theorem 5.4).

Example 5.12.[1] Let $N = 2$ and

$$T(\varkappa_1, \varkappa_2) = \begin{pmatrix} \varkappa_1 & \varkappa_2 \\ \varkappa_2 & -\varkappa_1 \end{pmatrix}. \tag{5.23}$$

[1] See RELLICH [1].

$T(\varkappa_1, \varkappa_2)$ is totally differentiable in $\varkappa_1, \varkappa_2$ and diagonable for all real values of $\varkappa_1, \varkappa_2$. But its eigenvalues

$$\lambda_{\pm}(\varkappa_1, \varkappa_2) = \pm(\varkappa_1^2 + \varkappa_2^2)^{1/2} \tag{5.24}$$

are not totally differentiable at $\varkappa_1 = \varkappa_2 = 0$.

We could also consider the case in which $T(\varkappa_1, \varkappa_2)$ is holomorphic in the two variables. But the eigenvalues of $T(\varkappa_1, \varkappa_2)$ might have rather complicated singularities, as is seen from the above example[1].

Remark 5.13. (5.23) is symmetric for real $\varkappa_1, \varkappa_2$ if the usual inner product is introduced into $\mathsf{X} = \mathrm{C}^2$. Thus the appearence of a singularity of the kind (5.24) shows that the situation is different from the case of a single variable, where the eigenvalue is holomorphic at $\varkappa = 0$ if $T(\varkappa)$ is normal for real $\varkappa$ (see Theorem 1.10).

Similar remarks apply to the case in which there are more than two variables.

We have seen above that in nonanalytic perturbations, the eigenprojections $P_h(\varkappa)$ and the eigennilpotents $D_h(\varkappa)$ may be quite ill-behaved. In most cases these troubles arise from the fact that the number of distinct eigenvalues can vary discontinuously with $\varkappa$. If we *assume* that the number s is constant in a domain D of $\varkappa$, the troubles usually disappear. We shall discuss some of the relevant results.

Theorem 5.13a[2]. *Let* $T(\varkappa) \in \mathscr{B}(\mathsf{X})$ *be* C^k *(k-times continuously differentiable) in* $\varkappa = (\varkappa_1, \ldots, \varkappa_m)$ *varying on a simply-connected domain* $\mathrm{D} \subset \mathrm{R}^m$. *Assume that the numbers s of distinct eigenvalues* $\lambda_h(\varkappa)$ *of* $T(\varkappa)$ *is constant for* $\varkappa \in \mathrm{D}$. *Then the* $\lambda_h(\varkappa)$ *are* C^k *in* D, *as well as the associated* $P_h(\varkappa)$ and $D_h(\varkappa)$.

Proof. Note that the numbering λ_h, P_h and D_h $(h = 1, \ldots, s)$ as indicated is possible throughout $\varkappa \in \mathrm{D}$, due to the continuity of the $\lambda_h(\varkappa)$ given in par. 1, 2. Since the assertions of the theorem are local properties in $\varkappa$, we may restrict $\varkappa$ to a small neighborhood of $\varkappa = 0$ (assuming $0 \in \mathrm{D}$, without loss of generality). Then $P_h(\varkappa)$ is given by the contour integral (1.16), in which $\Gamma = \Gamma_h$ may be chosen as a fixed circle about $\lambda_h(0)$ as long as $|\varkappa|$ is sufficiently small. Since $T(\varkappa)$ is C^k in $\varkappa$, it is easy to see that $R(\zeta, \varkappa)$ is also C^k in $\varkappa$ uniformly for $\zeta \in \Gamma_h$. It follows from (1.16) that $P_h(\varkappa)$ is C^k. Then formula (2.5) with $m = m_h = \dim P_h(\varkappa)$ $=$ const shows that $\lambda_h(\varkappa)$ is C^k, and (2.7) shows that the same is true with $D_h(\varkappa)$.

Remark 5.13b. Similar results hold when the C^k-property is replaced by analyticity in several variables $\varkappa = (\varkappa_1, \ldots, \varkappa_m)$ (real or complex), provided that s is assumed to be constant.

---

[1] But simple eigenvalues and the associated eigenprojections are again holomorphic in $\varkappa_1, \varkappa_2$; see Remark 5.13b.

[2] Cf. Nomizu [1].

Theorem 5.13c. *Let* $T(\varkappa) \in \mathscr{B}(\mathsf{X}_1, \mathsf{X}_2)$ *be* C^k *in* $\varkappa \in \mathrm{D}$, *where* $\mathrm{D} \subset \mathrm{R}^m$ *is as in Theorem* 5.13a. *Assume that* rank $T(\varkappa) = r$ *is constant for* $\varkappa \in \mathrm{D}$. *Then there are two families* $Q_j(\varkappa) \in \mathscr{B}(\mathsf{X}_j)$ *of projections* $(j = 1, 2)$ *of class* C^k *on* D *with* dim $Q_j(\varkappa) = r$, *such that* $T(\varkappa)$ *has range* $Q_2(\varkappa)\,\mathsf{X}_2$ *and null space* $(1 - Q_1(\varkappa))\,\mathsf{X}_1$. *There is an operator* $S(\varkappa) \in \mathscr{B}(\mathsf{X}_2, \mathsf{X}_1)$ *of class* C^k *in* $\varkappa \in \mathrm{D}$, *with range* $Q_1(\varkappa)\,\mathsf{X}_1$ *and null space* $(1 - Q_2(\varkappa))\,\mathsf{X}_2$, *such that* $S(\varkappa)\,T(\varkappa) = Q_1(\varkappa)$ *and* $T(\varkappa)\,S(\varkappa) = Q_2(\varkappa)$. *The* $Q_j(\varkappa)$ *may be chosen to be orthogonal projections if the* X_j *are made into Hilbert spaces by a suitable choice of the norm.*

Proof. We may assume that the X_j are Hilbert spaces and identify X_j^* with X_j. Then $T(\varkappa)^* \in \mathscr{B}(\mathsf{X}_2^*, \mathsf{X}_1^*) = \mathscr{B}(\mathsf{X}_2, \mathsf{X}_1)$ is also C^k in $\varkappa \in \mathrm{D}$. Let $H(\varkappa) = T(\varkappa)\,T(\varkappa)^*$. $H(\varkappa)$ is a nonnegative-definite, selfadjoint operator in X_2. According to I-§ 6.4, $H(\varkappa)$ has the same null space as $T(\varkappa)^*$ so that rank $H(\varkappa) = r$ and nul $H(\varkappa) = N - r$. Thus the eigenvalue zero of $H(\varkappa)$ has constant multiplicity $N - r$ and is separated from the other eigenvalues. The associated (orthogonal) eigenprojection $P(\varkappa)$ of dimension $N - r$ is C^k in $\varkappa$, as is seen by the argument given in the proof of Theorem 5.13a. Since $T(\varkappa)^*$ has the same null space $P(\varkappa)\,\mathsf{X}_2$ as $H(\varkappa)$, $T(\varkappa)$ has range $Q_2(\varkappa)\,\mathsf{X}_2$, where $Q_2(\varkappa) = 1 - P(\varkappa)$ with dimension r, and $Q_2(\varkappa)$ is also C^k in $\varkappa$.

Similarly we can prove, by considering the selfadjoint operator $T(\varkappa)^*\,T(\varkappa)$ in X_1, that $T(\varkappa)$ has null space $(1 - Q_1(\varkappa))\,\mathsf{X}_1$, where $Q_1(\varkappa)$ is an orthogonal projection of dimension r and C^k in $\varkappa$.

The relative inverse $S(\varkappa)$ can be constructed by

$$S(\varkappa) = T(\varkappa)^*\,K(\varkappa), \tag{5.24a}$$

where $K(\varkappa)$ is the reduced resolvement of $H(\varkappa)$ for the eigenvalue zero (see I-§ 5.3). As is easily verified, $K(\varkappa)$ is given by

$$K(\varkappa) = \frac{1}{2\pi i}\int_\Gamma (H(\varkappa) - \zeta)^{-1}\,\zeta^{-1}\,d\zeta, \tag{5.24b}$$

where Γ is a small circle about zero excluding all nonzero eigenvalues of $H(\varkappa)$. (5.24b) shows that $K(\varkappa)$ is C^k in $\varkappa$. From (5.24a) we have $T(\varkappa)\,S(\varkappa) = H(\varkappa)\,K(\varkappa) = 1 - P(\varkappa) = Q_2(\varkappa)$. Then $T(\varkappa)\,S(\varkappa)\,T(\varkappa) = Q_2(\varkappa)\,T(\varkappa) = T(\varkappa) = T(\varkappa)\,Q_1(\varkappa)$. Hence $T(\varkappa)\,[S(\varkappa)\,T(\varkappa) - Q_1(\varkappa)] = 0$. Since $S(\varkappa)\,\mathsf{X}_2 \subset T(\varkappa)^*\,\mathsf{X}_2 = Q_1(\varkappa)\,\mathsf{X}_1$, this gives $S(\varkappa)\,T(\varkappa) = Q_1(\varkappa)$, completing the proof of Theorem 5.13c.

Remark 5.13d. Let $m = 1$ in Theorem 5.13c. Then $Q_j(\varkappa)$ has an associated transformation function $U_j(\varkappa)$, which is also C^k in $\varkappa$; for $k = 1$ this was noted in Remark 4.3, and the same proof applies for general k. Moreover, the $U_j(\varkappa)$ are *unitary* for $\varkappa \in \mathrm{D}$ because the $Q_j(\varkappa)$ are orthogonal projections (see § 6.2 below). With these $U_j(\varkappa)$, we have the relation $T_0(\varkappa) = Q_2(0)\,T_0(\varkappa) = T_0(\varkappa)\,Q_1(0)$ for $T_0(\varkappa) = U_2(\varkappa)^{-1}\,T(\varkappa)\,U_1(\varkappa)$.

In other words, suitable unitary transformations $U_j(\varkappa)$ of class C^k in $\varkappa$ bring $T(\varkappa)$ into an operator $T_0(\varkappa)$ with a fixed range $Q_2(0)\,\mathsf{X}_2$ and a fixed null space $(1 - Q_1(0))\,\mathsf{X}_1$[1]. (cf. Remark 4.10 for the analytic case.)

8. The eigenvalues as functions of the operator

In perturbation theory the introduction of the parameter $\varkappa$ is sometimes rather artificial, although it sometimes corresponds to the real situation. We could rather consider the change of the eigenvalues of an operator T when T is changed by a small amount, without introducing any parameter $\varkappa$ or parameters $\varkappa_1, \varkappa_2, \ldots$. From this broader point of view, the eigenvalues of T should be regarded as functions of T itself. Some care is necessary in this interpretation, however, since the eigenvalues are not fixed in number. Again it is convenient to consider the unordered N-tuple $\mathfrak{S}[T]$, consisting of the N repeated eigenvalues of T, as a function of T. This is equivalent to regarding $\mathfrak{S}[T]$ as a function of the N^2 elements of the matrix representing T with respect to a fixed basis of X.

Theorem 5.14. *$\mathfrak{S}[T]$ is a continuous function of T. By this it is meant that, for any fixed operator T, the distance between $\mathfrak{S}[T + A]$ and $\mathfrak{S}[T]$ tends to zero for $\|A\| \to 0$.*

The proof of this theorem is contained in the result of par. 1,2 where the continuity of $\mathfrak{S}(\varkappa)$ as a function of $\varkappa$ is proved. An examination of the arguments given there will show that the use of the parameter $\varkappa$ is not essential.

This continuity of $\mathfrak{S}[T]$ is naturally *uniform* on any bounded region of the variable T (that is, a region in which $\|T\|$ is bounded), for the variable T is equivalent to N^2 complex variables as noted above. But the degree of continuity may be very weak at some special T (non-diagonable T), as is seen from the fact that the Puiseux series for the eigenvalues of $T + \varkappa T^{(1)} + \cdots$ can have the form $\lambda + \alpha \varkappa^{1/p} + \cdots$ [see (1.7) and Example 1.1, d)].

Let us now consider the differentiability of $\mathfrak{S}[T]$. As we have seen, the eigenvalues are not always differentiable even in the analytic case $T(\varkappa)$. If T is diagonable, on the other hand, the eigenvalues of $T + \varkappa T^{(1)}$ are differentiable at $\varkappa = 0$ for any $T^{(1)}$ (in the sense of par. 4), and the diagonability of T is necessary in order that this be true for every $T^{(1)}$. This proves

Theorem 5.15. *$\mathfrak{S}[T]$ is partially differentiable at $T = T_0$ if and only if T_0 is diagonable.*

---

[1] These results, which are variants of the corresponding ones for holomorphic families (Remark 4.10), are known as Doležal's theorem. See Doležal [1], Weiss and Falb [1]. Apparently it is not known if they are valid for $m > 1$.

Here "partially differentiable" means that $\mathfrak{S}[T + \varkappa T^{(1)}]$ is differentiable at $\varkappa = 0$ for any fixed $T^{(1)}$, and it implies the partial differentiability of $\mathfrak{S}[T]$ in each of the N^2 variables when it is regarded as a function of the N^2 matrix elements.

Theorem 5.15 is not true if "partially" is replaced by "totally". This is seen from Example 5.12, which shows that $\mathfrak{S}[T]$ need not be totally differentiable even when the change of T is restricted to a two-dimensional subspace of $\mathscr{B}(\mathsf{X})$. In general a complex-valued function $\mu[T]$ of $T \in \mathscr{B}(\mathsf{X})$ is said to be totally differentiable at $T = T_0$ if there is a function $\nu_{T_0}[A]$, *linear* in $A \in \mathscr{B}(\mathsf{X})$, such that

$$(5.25) \quad \|A\|^{-1} |\mu[T_0 + A] - \mu[T_0] - \nu_{T_0}[A]| \to 0 \quad \text{for} \quad \|A\| \to 0 .$$

This definition does not depend on the particular norm used, for all norms are equivalent. $\nu_{T_0}[A]$ is the *total differential* of $\mu[T]$ at $T = T_0$. It is easily seen that $\mu[T]$ is totally differentiable if and only if it is totally differentiable as a function of the N^2 matrix elements of T.

In reality we are here not considering a single complex-valued function $\mu[T]$ but an unordered N-tuple $\mathfrak{S}[T]$ as a function of T. If $\mathfrak{S}[T]$ were an *ordered* N-tuple, the above definition could be extended immediately to $\mathfrak{S}[T]$. But as $\mathfrak{S}[T]$ is unordered, this is not an easy matter and we shall not pursue it in this much generality. We shall rather restrict ourselves to the case in which T_0 is not only diagonable but simple (has N distinct eigenvalues). Then the same is true of $T = T_0 + A$ for sufficiently small $\|A\|$ in virtue of the continuity of $\mathfrak{S}[T]$, and the eigenvalues of T can be expressed in a neighborhood of T_0 by N single-valued, continuous functions $\lambda_h[T]$, $h = 1, \ldots, N$. We shall now prove

Theorem 5.16. *The functions $\lambda_h[T]$ are not only totally differentiable but holomorphic in a neighborhood of $T = T_0$.*

Remark 5.17. A complex-valued function $\mu[T]$ of T is said to be holomorphic at $T = T_0$ if it can be expanded into an absolutely convergent *power series* (Taylor series) in $A = T - T_0$:

$$(5.26) \quad \mu[T_0 + A] = \mu[T_0] + \mu^{(1)}[T_0, A] + \mu^{(2)}[T_0, A] + \cdots$$

in which $\mu^{(n)}[T_0, A]$ is a form of degree n in A, that is,

$$(5.27) \qquad \mu^{(n)}[T_0, A] = \mu^{(n)}[T_0; A, \ldots, A]$$

where $\mu^{(n)}[T_0; A_1, \ldots, A_n]$ is a symmetric n-linear form[1] in n operators

---

[1] A function $f(A_1, \ldots, A_n)$ is *symmetric* if its value is unchanged under any permutation of $A_1, \ldots, A_n$. It is n-linear if it is linear in each variable A_k.

$A_1, \ldots, A_n$. As is easily seen, $\mu[T]$ is holomorphic at $T = T_0$ if and only if $\mu[T_0 + A]$ can be expressed as a convergent power series in the N^2 matrix elements of A. In the same way holomorphic dependence of an operator-valued function $R[T] \in \mathscr{B}(\mathsf{X})$ on $T \in \mathscr{B}(\mathsf{X})$ can be defined.

Proof of Theorem 5.16. First we show that the one-dimensional eigenprojection $P_h[T]$ for the eigenvalue $\lambda_h[T]$ is holomorphic in T. We have, as in (1.17),

$$P_h[T_0 + A] = -\frac{1}{2\pi i} \int_{\Gamma_h} \sum_{n=0}^{\infty} R_0(\zeta) \, (-A\, R_0(\zeta))^n \, d\zeta \,, \tag{5.28}$$

where $R_0(\zeta) = (T_0 - \zeta)^{-1}$ and Γ_h is a small circle around $\lambda_h[T_0]$. The series in the integrand of (5.28) is uniformly convergent for $\zeta \in \Gamma_h$ for $\|A\| < \delta_h$, where δ_h is the minimum of $\|R_0(\zeta)\|^{-1}$ for $\zeta \in \Gamma_h$. Since the right member turns out to be a power series in A, we see that $P_h[T]$ is holomorphic at $T = T_0$.

Since $P_h[T]$ is one-dimensional, we have

$$\lambda_h[T_0 + A] = \operatorname{tr}\{(T_0 + A) \, P_h[T_0 + A]\} \,. \tag{5.29}$$

Substitution of the power series (5.28) shows that $\lambda_h[T_0 + A]$ is also a power series in A, as we wished to show.

§ 6. Perturbation of symmetric operators

1. Analytic perturbation of symmetric operators

Many theorems of the preceding sections can be simplified or strengthened if X is a unitary space H. For the reason stated at the beginning of I-§ 6., we shall mainly be concerned with the perturbation of *symmetric* operators.

Suppose we are given an operator $T(\varkappa)$ of the form (1.2) in which $T, T^{(1)}, T^{(2)}, \ldots$ are all symmetric. Then the sum $T(\varkappa)$ is also symmetric for *real* $\varkappa$. Naturally it cannot be expected that $T(\varkappa)$ be symmetric for all $\varkappa$ of a domain of the complex plane.

More generally, let us assume that we are given an operator-valued function $T(\varkappa) \in \mathscr{B}(\mathsf{H})$ which is holomorphic in a domain D_0 of the $\varkappa$-plane intersecting with the real axis and which is symmetric for real $\varkappa$:

$$T(\varkappa)^* = T(\varkappa) \quad \text{for real } \varkappa \,. \tag{6.1}$$

For brevity the family $\{T(\varkappa)\}$ will then be said to be *symmetric*. Also we shall speak of a symmetric perturbation when we consider $T(\varkappa)$ as a perturbed operator. $T(\bar{\varkappa})^*$ is holomorphic for $\varkappa \in \bar{D}_0$ (the mirror image of D_0 with respect to the real axis) and coincides with $T(\varkappa)$ for real $\varkappa$. Hence $T(\bar{\varkappa})^* = T(\varkappa)$ for $\varkappa \in D_0 \cap \bar{D}_0$ by the unique continuation property of holomorphic functions. Thus

$$T(\varkappa)^* = T(\bar{\varkappa}) \tag{6.2}$$

as long as both $\varkappa$ and $\bar{\varkappa}$ belong to D_0. This can be used to continue $T(\varkappa)$ analytically to any $\varkappa$ for which one of $\varkappa$ and $\bar{\varkappa}$ belongs to D_0 but the other does not. Thus we may assume without loss of generality that D_0 is symmetric with respect to the real axis.

Since a symmetric operator is normal, the following theorem results directly from Theorem 1.10.

Theorem 6.1. *If the holomorphic family $T(\varkappa)$ is symmetric, the eigenvalues $\lambda_h(\varkappa)$ and the eigenprojections $P_h(\varkappa)$ are holomorphic on the real axis, whereas the eigennilpotents $D_h(\varkappa)$ vanish identically*[1].

Problem 6.2. If $T(\varkappa) = T + \varkappa T^{(1)}$ with T and $T^{(1)}$ symmetric, the smallest eigenvalue of $T(\varkappa)$ for real $\varkappa$ is a piecewise holomorphic, concave function of $\varkappa$. [hint: Apply I-(6.79)].

Remark 6.3. Theorem 6.1 cannot be extended to the case of two or more variables. The eigenvalues of a function $T(\varkappa_1, \varkappa_2)$ holomorphic in $\varkappa_1, \varkappa_2$ and symmetric for real $\varkappa_1, \varkappa_2$ need not be holomorphic for real $\varkappa_1, \varkappa_2$, as is seen from Example 5.12.

Remark 6.4. A theorem similar to Theorem 6.1 holds if $T(\varkappa)$ is normal for real $\varkappa$ or, more generally, for all $\varkappa$ on a curve in D_0. But such a theorem is of little practical use, since it is not easy to express the condition in terms of the coefficients $T^{(n)}$ of (1.2).

The calculation of the perturbation series given in § 2 is also simplified in the case of a symmetric perturbation. Since the unperturbed operator T is symmetric, any eigenvalue λ of T is semisimple ($D = 0$) and the reduction process described in § 2.3 is effective. The operator function $\tilde{T}^{(1)}(\varkappa)$ given by (2.37) is again symmetric, for $P(\varkappa)$ is symmetric and commutes with $T(\varkappa)$. *The reduction process preserves symmetry*. Therefore, the reduction can be continued indefinitely. The splitting of the eigenvalues must come to an end after finitely many steps, however, and the eigenvalues and the eigenprojections are finally given explicitly by the formulas corresponding to (2.5) and (2.3) in the final stage of the reduction process where the splitting has ceased to occur. In this way *the reduction process gives a complete recipe for calculating explicitly the eigenvalues and eigenprojections in the case of a symmetric perturbation.*

Remark 6.5. Again there is no general criterion for deciding whether there is no further splitting of the eigenvalue at a given stage. But the reducibility principle given in Remark 2.4 is useful, especially for symmetric perturbations. Since the unperturbed eigenvalue at each stage is automatically semisimple, there can be no further splitting if the un-

[1] See Remark 1.11, however.

perturbed eigenprojection at that stage is irreducible with respect to a set $\{A\}$ of operators.

In applications such a set $\{A\}$ is often given as a *unitary group* under which $T(\varkappa)$ is invariant. Since the eigenprojection under consideration is an orthogonal projection, it is irreducible under $\{A\}$ if and only if there is no proper subspace of the eigenspace which is invariant under all the unitary operators A.

Remark 6.6. The general theory is simplified to some extent even if only the unperturbed operator T is symmetric or even normal. For example, all the eigenvalues $\lambda_h(\varkappa)$ are then continuously differentiable at $\varkappa = 0$ in virtue of the diagonability of T (Theorem 2.3). The estimates for the convergence radii and the error estimates are also simplified if the unperturbed operator T is symmetric or normal, as has been shown in § 3.5.

Remark 6.7. The estimate (3.52) is the best possible of its kind even in the special case of a symmetric perturbation, for Example 3.10 belongs to this case.

2. Orthonormal families of eigenvectors

Consider a holomorphic, symmetric family $T(\varkappa)$. For each real $\varkappa$ there exists an orthonormal basis $\{\varphi_n(\varkappa)\}$ of H consisting of eigenvectors of $T(\varkappa)$ [see I-(6.68)]. The question arises *whether these orthonormal eigenvectors $\varphi_n(\varkappa)$ can be chosen as holomorphic functions of $\varkappa$*. The answer is yes for real $\varkappa$.

Since the eigenvalues $\lambda_h(\varkappa)$ and the eigenprojections $P_h(\varkappa)$ are holomorphic on the real axis (Theorem 6.1), the method of § 4.5 can be applied to construct a holomorphic transformation function $U(\varkappa)$ satisfying (4.30). Furthermore, *$U(\varkappa)$ is unitary for real $\varkappa$*. To see this we recall that $U(\varkappa)$ was constructed as the solution of the differential equation $U' = Q(\varkappa)\, U$ with the initial condition $U(0) = 1$, where $Q(\varkappa)$ is given by (4.33). Since the $P_h(\varkappa)$ are symmetric, we have $P_h(\varkappa)^* = P_h(\bar{\varkappa})$ as in (6.2) and so the same is true of $P_h'(\varkappa)$. Hence $Q(\varkappa)$ is skew-symmetric: $Q(\bar{\varkappa})^* = -Q(\varkappa)$ and $U(\bar{\varkappa})^*$ satisfies the differential equation

$$\frac{d}{d\varkappa}\, U(\bar{\varkappa})^* = -\, U(\bar{\varkappa})^*\, Q(\varkappa)\,. \tag{6.3}$$

On the other hand, $V(\varkappa) = U(\varkappa)^{-1}$ satisfies the differential equation $V' = -\, V Q(\varkappa)$ with the initial condition $V(0) = 1$. In view of the uniqueness of the solution we must have

$$U(\bar{\varkappa})^* = U(\varkappa)^{-1}\,. \tag{6.4}$$

This shows that $U(\varkappa)$ is unitary for real $\varkappa$.

It follows that the basis $\varphi_{hk}(\varkappa) = U(\varkappa)\, \varphi_{hk}$ as given by (4.31) is orthonormal for real $\varkappa$ if the φ_{hk} form an orthonormal basis (which is possible since T is symmetric). It should be noted that the $\varphi_{hk}(\varkappa)$ are (not only generalized but proper) eigenvectors of $T(\varkappa)$ because $T(\varkappa)$ is diagonable. The existence of such an orthonormal basis depending smoothly on $\varkappa$ is one of the most remarkable results of the analytic perturbation theory for symmetric operators. That the analyticity is essential here will be seen below.

3. Continuity and differentiability

Let us now consider non-analytic perturbations of operators in H. Let $T(\varkappa) \in \mathscr{B}(\mathsf{H})$ depend continuously on the parameter $\varkappa$, which will now be assumed to be real. The eigenvalues of $T(\varkappa)$ then depend on $\varkappa$ continuously, and it is possible to construct N continuous functions $\mu_n(\varkappa)$, $n = 1, \ldots, N$, representing the repeated eigenvalues of $T(\varkappa)$ (see § 5.2). In this respect there is nothing new in the special case where $T(\varkappa)$ is symmetric, except that all $\mu_n(\varkappa)$ are real-valued and so a simple numbering such as (5.4) could also be used.

A new result is obtained for the differentiability of eigenvalues.

Theorem 6.8.[1] *Assume that $T(\varkappa)$ is symmetric and continuously differentiable in an interval* I *of $\varkappa$. Then there exist N continuously differentiable functions $\mu_n(\varkappa)$ on* I *that represent the repeated eigenvalues of $T(\varkappa)$.*

Proof. The proof of this theorem is rather complicated[2]. Consider a fixed value of $\varkappa$; we may set $\varkappa = 0$ without loss of generality. Let λ be one of the eigenvalues of $T = T(0)$, m its multiplicity, and P the associated eigenprojection. Since λ is semisimple, the derivatives at $\varkappa = 0$ of the repeated eigenvalues of $T(\varkappa)$ belonging to the λ-group are given by the m repeated eigenvalues of $P T'(0) P$ in the subspace $\mathsf{M} = P\mathsf{X}$ (Theorem 5.4). Let $\lambda'_1, \ldots, \lambda'_p$ be the *distinct* eigenvalues of $P T'(0) P$ in M and let $P_1, \ldots, P_p$ be the associated eigenprojections. The $\mathsf{M}_j = P_j \mathsf{H}$ are subspaces of M. It follows from the above remark that the λ-group (repeated) eigenvalues of $T(\varkappa)$ for small $|\varkappa| \neq 0$ are divided into p subgroups, namely the $\lambda + \varkappa\,\lambda'_j$-groups, $j = 1, \ldots, p$. Since each of these subgroups is separated from other eigenvalues, the total projections $P_j(\varkappa)$ for them are defined. $P_j(\varkappa)$ is at the same time the total projection for the λ'_j-group of the operator $\tilde{T}^{(1)}(\varkappa)$ given by (5.14). But $\tilde{T}^{(1)}(\varkappa)$ is continuous in a neighborhood of $\varkappa = 0$ as was shown there (the continuity for $\varkappa \neq 0$ is obvious). Hence $P_j(\varkappa)$ is continuous in a neighborhood of $\varkappa = 0$ by Theorem 5.1, and the same is true of

$$T_j(\varkappa) = P_j(\varkappa)\, T'(\varkappa)\, P_j(\varkappa) \tag{6.5}$$

because $T'(\varkappa) = dT(\varkappa)/d\varkappa$ is continuous by hypothesis.

---

[1] This theorem is due to RELLICH [8]. Recall that the result of this theorem is not necessarily true for a general (non-symmetric) perturbation (see Remark 5.8).

[2] The original proof due to RELLICH is even longer.

The $\lambda + \varkappa \lambda_j'$-group of $T(\varkappa)$ consists in general of several distinct eigenvalues, the number of which may change discontinuously with $\varkappa$ in any neighborhood of $\varkappa = 0$. Let $\lambda(\varkappa_0)$ be one of them for $\varkappa = \varkappa_0 \neq 0$ and let $Q(\varkappa_0)$ be the associated eigenprojection. This $\lambda(\varkappa_0)$ may further split for small $|\varkappa - \varkappa_0| \neq 0$, but the derivative of any of the resulting eigenvalues must be an eigenvalue of $Q(\varkappa_0)\, T'(\varkappa_0)\, Q(\varkappa_0)$ in the subspace $Q(\varkappa_0)\, \mathsf{H}$ (again by Theorem 5.4). But we have $Q(\varkappa_0)\,\mathsf{H} \subset P_j(\varkappa_0)\,\mathsf{H}$ because $\lambda(\varkappa_0)$ belongs to the $\lambda + \varkappa \lambda_j'$-group, so that $Q(\varkappa_0)\, T'(\varkappa_0)\, Q(\varkappa_0) = Q(\varkappa_0)\, T_j(\varkappa_0)\, Q(\varkappa_0)$. Therefore, the derivatives under consideration are eigenvalues of the orthogonal projection (in the sense of I-§ 6.10) of the operator $T_j(\varkappa_0)$ on a certain subspace of $\mathsf{M}_j(\varkappa_0) = P_j(\varkappa_0)\,\mathsf{H}$. Since $T_j(\varkappa_0)$ is symmetric, it follows from Theorem I-6.46 that these eigenvalues lie between the largest and the smallest eigenvalues of $T_j(\varkappa_0)$ in the subspace $\mathsf{M}_j(\varkappa_0)$. But as $T_j(\varkappa)$ is continuous in $\varkappa$ as shown above, the eigenvalues of $T_j(\varkappa_0)$ in $\mathsf{M}_j(\varkappa_0)$ tend for $\varkappa_0 \to 0$ to the eigenvalues of $P_j\, T'(0)\, P_j$ in M_j, which are all equal to λ_j'. It follows that the derivatives of the $\lambda + \varkappa \lambda_j'$-group eigenvalues of $T(\varkappa)$ must also tend to λ_j' for $\varkappa \to 0$. This proves the required continuity of the derivatives of the eigenvalues $\mu_n(\varkappa)$ constructed by Theorem 5.6. [In the above proof, essential use has been made of the symmetry of $T(\varkappa)$ in the application of Theorem I-6.46. This explains why the same result does not necessarily hold in the general case].

Remark 6.9. As in the general case, the eigenprojections or eigenvectors have much less continuity than the eigenvalues even in the case of a symmetric perturbation, once the assumption of analyticity is removed. Example 5.3 is sufficient to illustrate this; here the function $T(\varkappa)$ is infinitely differentiable in $\varkappa$ and symmetric (by the usual inner product in C^2), but it is impossible to find eigenvectors of $T(\varkappa)$ that are continuous at $\varkappa = 0$ and do not vanish there.

Remark 6.9a. Even for the eigenvalues, the situation is not much better in general. Theorem 6.8 is optimal in the sense that there is no corresponding result for higher differentiability. *Even if $T(\varkappa)$ is C^∞ in $\varkappa$, the $\mu_n(\varkappa)$ need not be C^2.* A counterexample is given by[1]

$$T(\varkappa) = \begin{pmatrix} e^{-(\alpha+\beta)/|\varkappa|} & e^{-\beta/|\varkappa|} \sin(1/|\varkappa|) \\ e^{-\beta/|\varkappa|} \sin(1/|\varkappa|) & -e^{-(\alpha+\beta)/|\varkappa|} \end{pmatrix}$$

for $\varkappa \neq 0$ and $T(0) = 0$, where α, β are positive constants. $T(\varkappa)$ is symmetric and C^∞ in $\varkappa \in R$. The eigenvalues of $T(\varkappa)$ are distinct for $\varkappa \neq 0$ and are given by

$$\mu_\pm(\varkappa) = \pm e^{-\beta/|\varkappa|} [e^{-2\alpha/|\varkappa|} + \sin^2(1/|\varkappa|)]^{1/2}$$

for $\varkappa \neq 0$, and $\mu_\pm(0) = 0$. Obviously $\mu_\pm(\varkappa)$ are C^∞ for $\varkappa \neq 0$, and it is easily verified that they are C^1 near $\varkappa = 0$ (thus confirming Theorem 6.8). But the second derivatives of $\mu_\pm(\varkappa)$ evaluated at $\varkappa = 1/n\pi$ $(n = 1, 2, \ldots)$ are $\pm(n\pi)^4 e^{(\alpha-\beta)n\pi} + O(e^{-n\pi\beta})$ as $n \to \infty$. Hence the second derivatives are discontinuous at $\varkappa = 0$ if $\alpha \geqq \beta$.

[1] Cf. Wasow [1].

4. The eigenvalues as functions of the symmetric operator

As in § 5.8, we can regard the eigenvalues of a symmetric operator T as functions of T itself. As before, the eigenvalues are continuous functions of T in the sense explained there. The situation is however much simpler now because we can, if we desire, regard the repeated eigenvalues as forming an *ordered* N-tuple by arranging them in the ascending order

$$\mu_1[T] \leqq \mu_2[T] \leqq \cdots \leqq \mu_N[T] . \tag{6.6}$$

This defines N real-valued functions of T, T varying over all symmetric operators in H. The continuity of the eigenvalues is expressed by the continuity of these functions ($T' \to T$ implies $\mu_n[T'] \to \mu_n[T]$).

The numbering (6.6) of the eigenvalues is very simple but is not always convenient, for the $\mu_n[T]$ are not necessarily even partially differentiable. This is seen, for example, by considering the $\mu_n[T + \varkappa T']$ as functions of $\varkappa$, where T, T' are symmetric and $\varkappa$ is real. The eigenvalues of $T + \varkappa T'$ can be represented as holomorphic functions of $\varkappa$ (Theorem 6.1). The graphs of these functions may cross each other at some values of $\varkappa$ (exceptional points). If such a *crossing* takes place, the graph of $\mu_n[T + \varkappa T']$ jumps from one smooth curve to another, making a corner at the crossing point. In other words, the $\mu_n[T + \varkappa T']$ are continuous but not necessarily differentiable. In any case, they are *piecewise holomorphic*, since there are only a finite number of crossing points (exceptional points) in any finite interval of $\varkappa$.

Thus it is sometimes more convenient to return to the old point of view of regarding the repeated eigenvalues of T as elements of an unordered N-tuple $\mathfrak{S}[T]$. Then it follows from the result of the preceding paragraph that *$\mathfrak{S}[T]$ is partially continuously differentiable*. But $\mathfrak{S}[T]$ is totally differentiable only at those T with N distinct eigenvalues (again Example 5.12). In the neighborhood of such a T, however, the functions (6.6) are not only differentiable but holomorphic in T.

5. Applications. A theorem of Lidskii

Perturbation theory is primarily interested in small changes of the various quantities involved. Here we shall consider some problems related to the change of the eigenvalues when the operator is subjected to a *finite* change[1]. More specifically, we consider the problem of estimating the relation between the eigenvalues of two symmetric operators A, B in terms of their difference $C = B - A$.

Let us denote respectively by α_n, β_n, γ_n, $n = 1, \ldots, N$, the repeated eigenvalues of A, B, C in the ascending order as in (6.6). Let

[1] For a more general study on finite changes of eigenvalues and eigenvectors, see Davis [1, 3], Davis and Kahan [1].

$$T(\varkappa) = A + \varkappa C\,, \quad 0 \leqq \varkappa \leqq 1\,, \tag{6.7}$$

so that $T(0) = A$ and $T(1) = B$, and denote by $\mu_n(\varkappa)$ the repeated eigenvalues of $T(\varkappa)$ in the ascending order. As shown in the preceding paragraph, the $\mu_n(\varkappa)$ are continuous and piecewise holomorphic, with $\mu_n(0) = \alpha_n$, $\mu_n(1) = \beta_n$. In the interval $0 \leqq \varkappa \leqq 1$ there are only a finite number of exceptional points where the derivatives of the $\mu_n(\varkappa)$ may be discontinuous.

According to par. 2, we can choose for each $\varkappa$ a complete orthonormal family $\{\varphi_n(\varkappa)\}$ consisting of eigenvectors of $T(\varkappa)$:

$$(T(\varkappa) - \mu_n(\varkappa))\,\varphi_n(\varkappa) = 0\,, \quad n = 1, \ldots, N\,, \tag{6.8}$$

in such a way that the $\varphi_n(\varkappa)$ are piecewise holomorphic. In general they are discontinuous at the exceptional points; this is due to the rather unnatural numbering of the eigenvalues $\mu_n(\varkappa)$. If $\varkappa$ is not an exceptional point, differentiation of (6.8) gives

$$(C - \mu_n'(\varkappa))\,\varphi_n(\varkappa) + (T(\varkappa) - \mu_n(\varkappa))\,\varphi_n'(\varkappa) = 0\,. \tag{6.9}$$

Taking the inner product of (6.9) with $\varphi_n(\varkappa)$ and making use of the symmetry of $T(\varkappa)$, (6.8) and the normalization $\|\varphi_n(\varkappa)\| = 1$, we obtain

$$\mu_n'(\varkappa) = (C\,\varphi_n(\varkappa)\,,\,\varphi_n(\varkappa))\,. \tag{6.10}$$

Since the $\mu_n(\varkappa)$ are continuous and the $\varphi_n(\varkappa)$ are piecewise continuous, integration of (6.10) yields

$$\beta_n - \alpha_n = \mu_n(1) - \mu_n(0) = \int_0^1 (C\,\varphi_n(\varkappa)\,,\,\varphi_n(\varkappa))\,d\varkappa\,. \tag{6.11}$$

Let $\{x_j\}$ be an orthonormal basis consisting of the eigenvectors of C:

$$C\,x_n = \gamma_n\,x_n\,, \quad n = 1, \ldots, N\,. \tag{6.12}$$

We have

$$(C\,\varphi_n(\varkappa), \varphi_n(\varkappa)) = \sum_j (C\,\varphi_n(\varkappa), x_j)\,(x_j, \varphi_n(\varkappa)) = \sum_j \gamma_j\,|(\varphi_n(\varkappa), x_j)|^2$$

and (6.11) becomes

$$\beta_n - \alpha_n = \sum \sigma_{nj}\,\gamma_j\,, \tag{6.13}$$

$$\sigma_{nj} = \int_0^1 |(\varphi_n(\varkappa), x_j)|^2\,d\varkappa\,. \tag{6.14}$$

The orthonormality of $\{\varphi_n(\varkappa)\}$ and $\{x_j\}$ implies that

$$\sum_j \sigma_{nj} = 1\,, \quad \sum_n \sigma_{nj} = 1\,, \quad \sigma_{nj} \geqq 0\,. \tag{6.15}$$

Now it is well known in matrix theory that a square matrix (σ_{nj}) with the properties (6.15) lies in the convex hull of the set of all permutation matrices[1]. Thus (6.13) leads to the following theorem due to LIDSKII [1].

Theorem 6.10. *Let A, B, C, α_n, β_n, γ_n be as above. The N-dimensional numerical vector $(\beta_1-\alpha_1, \ldots, \beta_N-\alpha_N)$ lies in the convex hull of the vectors obtained from $(\gamma_1, \ldots, \gamma_N)$ by all possible permutations of its elements.*

Another consequence of (6.13) is

Theorem 6.11. *For any convex function $\Phi(t)$ of a real variable t, the following inequality holds:*

$$\sum_n \Phi(\beta_n - \alpha_n) \leqq \sum_n \Phi(\gamma_n) . \tag{6.16}$$

The proof follows easily from (6.13), (6.15) and the convexity of Φ, for

$$\Phi(\beta_n - \alpha_n) = \Phi\left(\sum_j \sigma_{nj}\gamma_j\right) \leqq \sum_j \sigma_{nj}\Phi(\gamma_j) .$$

Example 6.12. Let $\Phi(t) = |t|^p$ with $p \geqq 1$. Then (6.16) gives[2]

$$\sum_n |\beta_n - \alpha_n|^p \leqq \sum_n |\gamma_n|^p , \quad p \geqq 1 . \tag{6.17}$$

Taking the p-th root of (6.17) and letting $p \to \infty$, we obtain

$$\max_n |\beta_n - \alpha_n| \leqq \max_n |\gamma_n| = \|C\|. \tag{6.18}$$

6. Nonsymmetric perturbation of symmetric operators

From (6.18) we see that the eigenvalues of a symmetric operator A does not change by more than $\|C\|$ when A is perturbed by a symmetric operator C (assuming that the eigenvalues are properly numbered).

A similar result holds for a nonsymmetric perturbation C, provided $\|C\|$ is sufficiently small and A is symmetric. Indeed, an eigenvalue λ of A does not change by more than $\|C\|$ if $\|C\| < r/2$, where r is the isolation distance of λ (Theorem 3.9). Note that λ need not be simple; if λ has multiplicity m, there are several eigenvalues of $A + C$ with total multiplicity m within the distance $r/2$ of λ. The same is true even when we consider a group of several eigenvalues of A simultaneously, if this group has a distance r from the rest of the spectrum of A.

Such a result need not hold if $\|C\|$ is large compared with the spacing of the eigenvalues of A. As a result, the distance between $\mathfrak{S}(A)$ and

[1] See BIRKHOFF [1], MIRSKY [1]. A permutation matrix (σ_{jk}) is associated with a permutation $j \to \pi(j)$ of $\{1, 2, \ldots, n\}$ by the relation $\sigma_{jk} = 1$ if $k = \pi(j)$ and $\sigma_{jk} = 0$ otherwise.

[2] It was shown by HOFFMAN and WIELANDT [1] that (6.17) is true for $p = 2$ in a more general case in which A, B are only assumed to be normal, if the right member is replaced by $\operatorname{tr} C^* C$ and if a suitable numbering of $\{\alpha_n\}$ is chosen. Note that $C = B - A$ need not be normal for normal A, B.

$\mathfrak{S}(A + C)$ may be arbitrarily large compared with $\|C\|$. The following example[1] illustrates the situation.

Example 6.13. Let $A = (a_{jk})$ be an $N \times N$ symmetric matrix with $a_{jk} = |j-k|^{-1}$ for $j \neq k$ and $a_{jj} = 0$ $(j, k = 1, \ldots, N)$. Let $C = (c_{jk})$ be a skew-symmetric matrix with $c_{jk} = (j - k)^{-1}$ for $j \neq k$ and $c_{jj} = 0$. Then $A + C$ is a lower triangular matrix with diagonal elements zero, so that it is nilpotent and $\Sigma(A + C) = \{0\}$. Now it can be shown that $\|C\| \leqq \pi$ for any $N = 1, 2, \ldots$, while $\|A\|$ grows with N logarithmically so that $\Sigma(A)$ has diameter proportional to $\log N$. Thus a spectrum of arbitrarily large diameter of a symmetric operator can be converted into a single point by a perturbation of a fixed size.

§7. Perturbation of (essentially) nonnegative matrices

1. Monotonicity of the principal eigenvalue

In this section we consider some basic results related to the perturbation of (essentially) nonnegative matrices introduced in I-§ 7. We recall that positivity involved here is different from the positive-definiteness for symmetric operators.

A matrix T is essentially nonnegative, by definition, if all the off-diagonal elements of T are nonnegative (I-§ 7.3). Such matrices form a closed subset $\mathcal{G}_+(\mathsf{X})$ of $\mathscr{B}(\mathsf{X})$, where $\mathsf{X} = \mathbf{C}^N$. For each $T \in \mathcal{G}_+(\mathsf{X})$, the principal eigenvalue $\lambda[T]$ is defined (Theorem I-7.5). $\lambda[T]$ is real, has a nonnegative eigenvector, and is strictly larger than the real part of any other eigenvalue of T. If T is irreducible, $\lambda[T]$ is simple and has a positive eigenvector (Theorem I-7.10). For simplicity we denote by $\mathcal{G}_{++}(\mathsf{X})$ the set of all irreducible, essentially nonnegative matrices.

Lemma 7.1. *$\mathcal{G}_{++}(\mathsf{X})$ is relatively open and dense in $\mathcal{G}_+(\mathsf{X})$. If $T \in \mathcal{G}_{++}(\mathsf{X})$ and $A \in \mathcal{G}_+(\mathsf{X})$, then $T + A \in \mathcal{G}_{++}(\mathsf{X})$.*

Proof. According to Theorem I-7.7 and the definition of irreducibility, $T \in \mathcal{G}_+(\mathsf{X})$ is in $\mathcal{G}_{++}(\mathsf{X})$ if and only if e^T is positive. This property is preserved when T is varied slightly. Hence $\mathcal{G}_{++}(\mathsf{X})$ is open in $\mathcal{G}_+(\mathsf{X})$. [Note that this argument is valid only under the assumption that $T \in \mathcal{G}_+(\mathsf{X})$.] Again, $T \in \mathcal{G}_{++}(\mathsf{X})$ if T is essentially positive. But essentially positive matrices are obviously dense in $\mathcal{G}_+(\mathsf{X})$. Hence $\mathcal{G}_{++}(\mathsf{X})$ is dense in $\mathcal{G}_+(\mathsf{X})$. The last statement of the lemma was proved in I-§ 7.4 (in the remark following the definition of irreducibility).

Theorem 7.2. *$\lambda[T]$ is a monotone nondecreasing function of $T \in \mathcal{G}_+(\mathsf{X})$. More precisely, if $T, S \in \mathcal{G}_+(\mathsf{X})$ with $S - T$ nonnegative, then $\lambda[T] \leqq \lambda[S]$. If in particular $T \in \mathcal{G}_{++}(\mathsf{X})$, equality holds only for $S = T$.*

Proof. Set $A = S - T$. We may assume that A is nonnegative but not zero. First assume that $T \in \mathcal{G}_{++}(\mathsf{X})$ and consider the family $T(\varkappa) = T + \varkappa A$, $0 \leqq \varkappa \leqq 1$. Then $T(\varkappa) \in \mathcal{G}_{++}(\mathsf{X})$ by Lemma 7.1, so that

[1] See KAHAN [1, 2].

$\lambda(\varkappa) = \lambda[T(\varkappa)]$ is a simple eigenvalue of $T(\varkappa)$. It follows from analytic perturbation theory (§ 1) that $\lambda(\varkappa)$ is holomorphic in $\varkappa$ in a complex neighborhood of the interval $[0, 1]$, and we have $\lambda'(0) = (A\varphi, \psi) > 0$, where $\varphi \gg 0$ and $\psi \gg 0$ are the eigenvectors of T and T^*, respectively, for the eigenvalue $\lambda(0) = \lambda[T] = \lambda[T^*]$ normalized so that $(\varphi, \psi) = 1$ (see (2.36)). Similarly we have $\lambda'(\varkappa) > 0$ for $0 \leqq \varkappa \leqq 1$, since $\varkappa = 0$ is not a distinguished value. Thus we have $\lambda[S] = \lambda(1) > \lambda(0) = \lambda[T]$, and the theorem is proved for $T \in \mathscr{G}_{++}(\mathsf{X})$. The general case $T \in \mathscr{G}_{+}(\mathsf{X})$ can be dealt with by continuity, by approximating T with irreducible matrices (see Lemma 7.1).

Example 7.3. (a) $T = \begin{pmatrix} 0 & 1 \\ 0 & 0 \end{pmatrix}$, $A = \begin{pmatrix} 0 & 1 \\ 0 & 0 \end{pmatrix}$, $S = \begin{pmatrix} 0 & 2 \\ 0 & 0 \end{pmatrix}$, $\lambda[T] = \lambda[S] = 0$. T is reducible.

(b) $T = \begin{pmatrix} 0 & 1 \\ 1 & 0 \end{pmatrix}$, $A = \begin{pmatrix} 0 & 1 \\ 0 & 0 \end{pmatrix}$, $S = \begin{pmatrix} 0 & 2 \\ 1 & 0 \end{pmatrix}$, $\lambda[T] = 1$, $\lambda[S] = 2^{1/2}$. T is irreducible.

2. Convexity of the principal eigenvalue

$\lambda[T]$ has a convexity property in a restricted sense: it is a convex function of the *diagonal elements* of T. This property may be stated, equivalently, in the following form.

Theorem 7.4 [1]. *Let $T \in \mathscr{G}_{+}(\mathsf{X})$ and let A be a real diagonal matrix. Then $T + \varkappa A \in \mathscr{G}_{+}(\mathsf{X})$ for $-\infty < \varkappa < \infty$, and $\lambda[T + \varkappa A]$ is a convex function of $\varkappa$. If, in particular, $T \in \mathscr{G}_{++}(\mathsf{X})$ and A is not a scalar multiple of the unit matrix, then $\lambda[T + \varkappa A]$ is strictly convex in $\varkappa$ in any interval.*

Remark 7.5. This restricted convexity of $\lambda[T]$ is an analog of the convexity of the largest eigenvalue $\lambda_N[H]$ of a *symmetric* operator H. If H and K are symmetric, $\lambda_N[H + \varkappa K]$ is a convex function of the real parameter $\varkappa$ [see Problem 6.2 and I-(6.79), where the results were stated for the smallest eigenvalue $\lambda_1[H]$, for which one has concavity rather than convexity].

Example 7.6. (a) In general there is no convexity of $\lambda[T]$ when the off-diagonal elements are varied. A counterexample is

$$T(\varkappa) = \begin{pmatrix} 1 & \varkappa \\ 1 & 1 \end{pmatrix}, \quad \lambda[T(\varkappa)] = 1 + \varkappa^{1/2}, \quad \varkappa \geqq 0.$$

$1 + \varkappa^{1/2}$ is not a convex function of $\varkappa \geqq 0$.

(b) For strict convexity, it is essential that $T \in \mathscr{G}_{++}(\mathsf{X})$. A counterexample is

[1] This theorem was proved by Cohen by different methods. A particularly simple proof is found in Cohen [1]. See also Friedland [1].

$T(\varkappa) = \begin{pmatrix} \varkappa & 1 \\ 0 & -\varkappa \end{pmatrix}$, $\lambda[T(\varkappa)] = |\varkappa|$. $T(\varkappa)$ is reducible. $|\varkappa|$ is not strictly convex in any interval of $\varkappa$.

Proof of Theorem 7.4. In view of Lemma 7.1, it suffices to prove the second part of the theorem (cf. the proof of Theorem 7.2). Thus we assume that $T \in \mathscr{G}_{++}(\mathsf{X})$ so that $T + \varkappa A \in \mathscr{G}_{++}(\mathsf{X})$ too for all $\varkappa$ (see Lemma 7.1). To prove the convexity of $\lambda(\varkappa) = \lambda[T + \varkappa A]$, it suffices to show that $\lambda''(0) \geqq 0$, since $\varkappa = 0$ is not a distinguished value.

To this end, we shall use the perturbation series (2.21). Before doing so, however, we transform the operator T into a convenient form. According to Theorem I-7.12, there is a diagonal matrix F with positive diagonal elements such that $F^{-1} T F - \lambda$ is dissipative, where $\lambda = \lambda(0) = \lambda[T]$. (Here $\mathsf{X} = \mathbf{C}^N$ is regarded as a unitary space, with $\mathsf{X}^* = \mathsf{X}$.) Set

$$T_0 = F^{-1} T F, \quad A_0 = F^{-1} A F. \tag{7.1}$$

Then $T_0 \in \mathscr{G}_{++}(\mathsf{X})$ (because $e^{T_0} = F^{-1} e^T F$ is positive with e^T) and A_0 is diagonal. Furthermore,

$$\lambda(\varkappa) = \lambda[T + \varkappa A] = \lambda[F^{-1}(T + \varkappa A) F] = \lambda[T_0 + \varkappa A_0], \tag{7.2}$$

since a similarity transformation does not change the eigenvalues. To prove that $\lambda''(0) \geqq 0$, therefore, we may assume that $T = T_0$, $A = A_0$.

Then $T - \lambda$ is dissipative. It follows from Theorem I-6.49 that the eigenvalue λ of T is semisimple (which we already know), the associated eigenprojection P is selfadjoint, and that the reduced resolvent S is dissipative. Thus

$$\operatorname{Re} \operatorname{tr}(ASAP) = \operatorname{Re}(SA\varphi, A\varphi) \leqq 0, \tag{7.3}$$

where $P = (\ \ , \varphi)\varphi$ and $\varphi \gg 0$ is the normalized eigenvector of T for the principal eigenvalue λ. In view of formula (2.33), which gives $\hat{\lambda}^{(2)} = 2\lambda''(0)$, it follows that $\lambda''(0) = \operatorname{Re}\lambda''(0) \geqq 0$.

Thus we have proved that $\lambda(\varkappa)$ is convex in $\varkappa$. To prove that it is strictly convex, it suffices to show that $\lambda(\varkappa)$ is not linear in $\varkappa$, since we know that it is real analytic for $-\infty < \varkappa < \infty$. (Here it is important that $\lambda(\varkappa)$ is a simple eigenvalue of $T + \varkappa A$.) To this end, let α' and α'' be, respectively, the smallest and the largest eigenvalue of A. Since $T + \varkappa A = \varkappa(A + \varkappa^{-1} T)$, we have

$$\lim_{\varkappa \to +\infty} \varkappa^{-1} \lambda[T + \varkappa A] = \alpha'', \quad \lim_{\varkappa \to -\infty} \varkappa^{-1} \lambda[T + \varkappa A] = \alpha', \tag{7.4}$$

by the continuity of the eigenvalues of $A + \varkappa^{-1} T$ as $\varkappa \to \pm\infty$ (see § 5.1). In other words, $\lambda[T + \varkappa A]$ is asymptotically equal to $\alpha''\varkappa$ as $\varkappa \to +\infty$ and to $\alpha'\varkappa$ as $\varkappa \to -\infty$. If A is not a scalar multiple of the unit matrix, $\alpha' < \alpha''$ and the analytic function $\lambda[T + \varkappa A]$ cannot be linear in $\varkappa$.

Bibliography

AKHIEZER, N. I. and I. M. GLAZMAN: ⟦1⟧ Theory of linear operators in Hilbert space (English translation), Vol. I and II, New York: Frederick Ungar 1961 and 1963.

BART, H.: [1] Holomorphic relative inverses of operator valued functions. Math. Ann. **208**, 179–194 (1974).

BAUMGÄRTEL, H.: ⟦1⟧ Endlichdimensionale analytische Störungstheorie. Berlin: Akademie-Verlag 1972. [1] Zur Störungstheorie beschränkter linearer Operatoren eines Banachschen Raumes. Math. Nachr. **26**, 361–379 (1964). [4] Analytische Störung isolierter Eigenwerte endlicher algebraischer Vielfachheit von nichtselbstadjungierten Operatoren. Monatsb. Deutsch. Akad. Wiss. Berlin **10**, 250–257 (1968). [5] Jordansche Normalform holomorpher Matrizen. Monatsb. Deutsch. Akad. Wiss. Berlin **11**, 23–24 (1969). [6] Ein Reduktionsprozeß für analytische Störungen nichthalbeinfacher Eigenwerte. Monatsb. Deutsch. Akad. Wiss. Berlin **11**, 81–89 (1969). [7] Zur Abschätzung der Konvergenzradien von Störungsreihen. Monatsb. Deutsch. Akad. Wiss. Berlin **11**, 556–572 (1973).

BIRKHOFF, G.: [1] Three observations on linear algebra. Univ. Nac. Tucumán Rev. Ser. A. **5**, 147–151 (1946).

BLOCH, C.: [1] Sur la théorie des perturbations des états liés. Nuclear Phys. **6**, 329–347 (1958).

BURCKEL, R. B.: ⟦1⟧ An introduction to classical complex analysis. Vol. 1. Basel and Stuttgart: Birkhäuser 1979.

BUTLER, J. B. JR.: [1] Perturbation series for eigenvalues of analytic non-symmetric operators. Arch. Math. **10**, 21–27 (1959).

COHEN, J. E.: [1] Convexity of the dominant eigenvalue of an essentially nonnegative matrix. Proc. Amer. Math. Soc. **81**, 657–658 (1981).

COURANT, R. and D. HILBERT: ⟦1⟧ Methods of mathematical physics, I. New York: Interscience 1953.

DAVIS, C.: [1] The rotation of eigenvectors by a perturbation. J. Math. Anal. Appl. **6**, 159–173 (1963). [2] Separation of two linear subspaces. Acta Sci. Math. Szeged. **19**, 172–187 (1958). [3] The rotation of eigenvectors by a perturbation, II. J. Math. Anal. Appl. **11**, 20–27 (1965).

DAVIS, C. and W. M. KAHAN: [1] The rotation of eigenvectors by a perturbation. III. SIAM J. Numer. Anal. **7**, 1–46 (1970).

DEL PASQUA, D.: [1] Su una nozioni di varietà lineari disgiunte di uno spazio di Banach. Rendi. di Mat. **13**, 1–17 (1955).

DOLEŽAL, V.: [1] The existence of a continuous basis of a certain linear subspace of E_r which depends on a parameter. Casopis Pest. Mat. **89**, 466–469 (1964).

DUNFORD, N. and J. T. SCHWARTZ: ⟦1⟧ Linear operators, Part I: General theory; Part II: Spectral theory; Part III: Spectral operators. New York: Interscience 1958, 1963, 1971.

EGGLESTON, H. G.: ⟦1⟧ Convexity. Cambridge: University Press 1963.

FRIEDLAND, S.: [1] Convex spectral functions, Linear and Multilinear Algebra **9**, 299–316 (1981).

FRIEDRICHS, K. O.: ⟦1⟧ Perturbation of spectra in Hilbert space. Providence: Amer. Math. Soc. 1965. [2] Über die Spektralzerlegung eines Integraloperators. Math. Ann. **115**, 249–272 (1938).

GANTMACHER, F. R.: ⟦1⟧ The theory of matrices (English translation) Vol. I, II. New York: Chelsea 1959.

GARRIDO, L. M.: [1] Generalized adiabatic invariance. J. Mathematical Phys. **5**, 355–362 (1964).

GARRIDO, L. M. and F. J. SANCHO: [1] Degree of approximate validity of the adiabatic invariance in quantum mechanics. Physica **28**, 553–560 (1962).

GELFAND, I. M.: ⟦1⟧ Lectures on linear algebra (English translation). New York: Interscience 1961.

GOHBERG, I. C. and M. G. KREIN: [1] The basic propositions on defect numbers, root numbers, and indices of linear operators. Uspehi Mat. Nauk 12, **2** (74), 43–118 (1957); Amer. Math. Soc. Translations Ser. 2, **13**, 185–264 (1960).

GRAMSCH, B.: [2] Inversion von Fredholmfunktionen bei stetiger und holomorpher Abhängigkeit von Parametern. Math. Ann. **214**, 95–147 (1975).

HALMOS, P. R.: ⟦2⟧ Finite-dimensional vector spaces, 2nd Ed. Princeton: D. van Nostrand 1958.

HARDY, G. H., J. E. LITTLEWOOD and G. PÓLYA: ⟦1⟧ Inequalities. 2nd Ed. Cambridge: University Press 1952.

HELMER, O.: [1] Divisibility properties of integral functions. Duke Math. J. **6**, 345–356 (1940).

HILLE, E. and R. S. PHILLIPS: ⟦1⟧ Functional analysis and semi-groups. Revised ed. Providence: Amer. Math. Soc. 1957.

HOFFMAN, K. and R. KUNZE: ⟦1⟧ Linear algebra. Englewood Cliffs: Prentice-Hall 1961.

HOFFMAN, A. J. and H. W. WIELANDT: [1] The variation of the spectrum of a normal matrix. Duke Math. J. **20**, 37–39 (1953).

KAHAN, W.: [1] Every $n \times n$ matrix Z with real spectrum satisfies $\|Z - Z^*\| \leqq \|Z + Z^*\| (\log_2 n + 0.038)$. Proc. Amer. Math. Soc. **39**, 235–241 (1973). [2] Spectra of nearly Hermitian matrices. Proc. Amer. Math. Soc. **48**, 11–17 (1975).

KATO, T.: [1] On the convergence of the perturbation method, I, II. Progr. Theor. Phys. **4**, 514–523 (1949); **5**, 95–101; 207–212 (1950). [2] On the adiabatic theorem of quantum mechanics. J. Phys. Soc. Japan **5**, 435–439 (1950). [3] On the convergence of the perturbation method. J. Fac. Sci. Univ. Tokyo Sect. I, **6**, 145–226 (1951). [6] On the perturbation theory of closed linear operators. J. Math. Soc. Japan **4**, 323–337 (1952). [9] Notes on projections and perturbation theory. Technical Report No. 9, Univ. Calif. 1955. [12] Perturbation theory for nullity, deficiency and other quantities of linear operators. J. Analyse Math. **6**, 261–322 (1958). [13] Estimation of iterated matrices, with application to the von Neumann condition. Numer. Math. **2**, 22–29 (1960).

KEMBLE, E. C.. ⟦1⟧ The fundamental principles of quantum mechanics. New York: Dover 1958.

KNOPP, K.: ⟦1, 2⟧ Theory of functions (English translation). Parts I and II. New York: Dover 1945 and 1947.

LIDSKII, V. B.: [1] The proper values of the sum and product of symmetric matrices. Dokl. Akad. Nauk SSSR **75**, 769–772 (1950) (Russian).

LIVŠIC, B. L.: [1] Perturbation theory for a simple structure operator. Dokl. Akad. Nauk SSSR **133**, 800–803 (1960) (Russian).

LORCH, E. R.: ⟦1⟧ Spectral theory. New York: Oxford University Press 1962.

MIRSKY, L.: [1] Proofs of two theorems on doubly-stochastic matrices. Proc. Amer. Math. Soc. **9**, 371–374 (1958).

MOTSKIN, T. S. and O. TAUSSKY: [1] Pairs of matrices with property L. Trans. Amer. Math. Soc. **73**, 108–114 (1952). [2] Pairs of matrices with property L. II. Trans. Amer. Math. Soc. **80**, 387–401 (1955).

NOMIZU, K.: [1] Characteristic roots and vectors of a differentiable family of symmetric matrices. Linear and Multilinear Algebra **1**, 159–162 (1973).

PARLETT, B. N.: [[1]] The symmetric eigenvalue problem. Englewood Cliffs: Prentice-Hall 1980.

PHILLIPS, R. S.: [1] Perturbation theory for semi-groups of linear operators. Trans. Amer. Math. Soc. **74**, 199–221 (1954).

PÓLYA, G. and G. SZEGÖ: [[1]] Aufgaben und Lehrsätze aus der Analysis, I. 3. Aufl. Berlin-Göttingen-Heidelberg: Springer 1964.

PORATH, G.: [1] Störungstheorie der isolierten Eigenwerte für abgeschlossene lineare Transformationen im Banachschen Raum. Math. Nachr. **20**, 175–230 (1959). [2] Störungstheorie für lineare Transformationen im Banachschen Raum. Wiss. Z. Tech. Hochsch. Dresden **9**, 1121–1125 (1959/60).

RAYLEIGH, LORD: [[1]] The theory of sound. Vol. I. London: 1927.

REED, M. and B. SIMON: [[2]] Methods of modern mathematical physics. Vol. IV. New York-London: Academic Press 1978.

RELLICH, F.: [1] Störungstheorie der Spectralzerlegung, I. Math. Ann. **113**, 600–619 (1937). [2] Störungstheorie der Spektralzerlegung, II. Math. Ann. **113**, 677–685 (1937). [3] Störungstheorie der Spektralzerlegung, III. Math. Ann. **116**, 555–570 (1939). [4] Störungstheorie der Spektralzerlegung, IV. Math. Ann. **117**, 356–382 (1940). [5] Störungstheorie der Spektralzerlegung, V. Math. Ann. **118**, 462–484 (1942). [6] Störungstheorie der Spektralzerlegung. Proc. Intern. Congress Math. 1950, **I**, 606–613. [7] New results in the perturbation theory of eigenvalue problems. Nat. Bur. Standards Appl. Math. Ser. **29**, 95–99 (1953). [8] Perturbation theory of eigenvalue problems. Lecture Notes, New York Univ. 1953.

RIESZ, F. and B. SZ.-NAGY: [[1]] Functional analysis (English translation). New York: Frederick Ungar 1955.

ROSENBLOOM, P.: [1] Perturbation of linear operators in Banach spaces. Arch. Math. **6**, 89–101 (1955).

ROYDEN, H. L.: [[1]] Real analysis. New York: Macmillan 1963.

SAPHAR, P.: [1] Contribution a l'étude des applications linéaires dans un espace de Banach, Bull. Soc. Math. France **92**, 363–384 (1964).

SCHAEFER, H. H.: [[1]] Banach lattices and positive operators. New York-Heidelberg-Berlin: Springer 1974.

SCHÄFKE, F. W.: [3] Über Eigenwertprobleme mit zwei Parametern. Math. Nachr. **6**, 109–124 (1951). [4] Verbesserte Konvergenz- und Fehlerabschätzungen für die Störungsrechnung. Z. angew. Math. Mech. **33**, 255–259 (1953). [5] Zur Störungstheorie der Spektralzerlegung. Math. Ann. **133**, 219–234 (1957).

SCHIFF, L. I.: [[1]] Quantum mechanics. New York-Toronto-London: McGraw-Hill 1955.

SCHRÖDER, J.: [1] Fehlerabschätzungen zur Störungsrechnung bei linearen Eigenwertproblemen mit Operatoren eines Hilbertschen Raumes. Math. Nachr. **10**, 113–128 (1953). [2] Fehlerabschätzungen zur Störungsrechnung für lineare Eigenwertprobleme bei gewöhnlichen Differentialgleichungen. Z. angew. Math. Mech. **34**, 140–149 (1954). [3] Störungsrechnung bei Eigenwert- und Verzweigungsaufgaben. Arch. Rational Mech. Anal. **1**, 436–468 (1958).

SCHRÖDINGER, E.: [[1]] Collected papers on wave mechanics. London and Glasgow: 1928. [1] Quantisierung als Eigenwertproblem. (Dritte Mitteilung: Störungstheorie, mit Anwendung auf den Starkeffekt der Balmerlinien.) Ann. Physik **80**, 437–490 (1926).

SILVERMAN, L. M. and R. S. BUCY: [1] Generalizations of a theorem of Doležal. Math. Systems Theory **4**, 334–339 (1970).

ŠMUL'YAN, YU. L.: [1] Completely continuous perturbation of operators. Dokl. Akad. Nauk SSSR **101**, 35–38 (1955) (Russian).

STONE, M. H.: [1] Linear transformations in Hilbert space and their applications to analysis. Providence: Amer. Math. Soc. 1932.

SZ.-NAGY, B.: [1] Perturbations des transformations autoadjointes dans l'espace de Hilbert. Comment. Math. Helv. **19**, 347–366 (1946/47). [2] Perturbations des transformations linéaires fermées. Acta Sci. Math. Szeged. **14**, 125–137 (1951).

TITCHMARSH, E. C.: [1] Some theorems on perturbation theory. Proc. Roy. Soc. London Ser. A, **200**, 34–46 (1949). [2] Some theorems on perturbation theory. II. Proc. Roy. Soc. London Ser. A, **201**, 473–479 (1950).

VIŠIK, M. I. and L. A. LYUSTERNIK: [1] Perturbation of eigenvalues and eigenelements for some non-selfadjoint operators. Dokl. Akad. Nauk SSSR **130**, 251–253 (1960) (Russian). [3] Solution of some perturbation problems in the case of matrices and self-adjoint or non-self-adjoint differential equations. I. Uspehi Mat. Nauk **15**, 3 (93), 3–80 (1960); Russian Math. Surveys 15, no. 3, 1–73 (1960).

WASOW, W.: [1] On the spectrum of Hermitian matrix-valued functions. Resultate der Math. **2**, 206–214 (1979).

WEISS, L. and P. L. FALB: [1] Doležal's theorem, linear algebra with continuously parametrized elements, and time-varying systems. Math. Systems Theory **3**, 67–75 (1969).

WILSON, A. H.: [1] Perturbation theory in quantum mechanics. I. Proc. Roy. Soc. London Ser. A, **122**, 589–598 (1929).

WOLF, F.: [1] Analytic perturbation of operators in Banach spaces. Math. Ann. **124**, 317–333 (1952).

Notation Index

Author index

Subject index